AFTER THE SANDS

ENERGY AND ECOLOGICAL SECURITY
FOR CANADIANS

GORDON LAXER

Douglas & McIntyre

Douglas and McIntyre (2013) Ltd.
P.O. Box 219, Madeira Park, BC, V0N 2H0
www.douglas-mcintyre.com

Edited by Silas White and Merrie-Ellen Wilcox
Cover design by Anna Comfort O'Keeffe
Cover Illustration by Dave Murray
Text design by Lisa Eng-Lodge, Electra Design Group
Printed and bound in Canada

Douglas and McIntyre (2013) Ltd. acknowledges the support of the Canada Council for the Arts, which last year invested $157 million to bring the arts to Canadians throughout the country. We also gratefully acknowledge financial support from the Government of Canada through the Canada Book Fund and from the Province of British Columbia through the BC Arts Council and the Book Publishing Tax Credit.

Cataloguing data available from Library and Archives Canada
ISBN 978-1-77162-100-7 (paper)
ISBN 978-1-77162-101-4 (ebook)

Advance Praise for *After the Sands*

This is a myth-destroying blockbuster book. Gordon Laxer takes apart convenient corporate narratives with hard facts about Canada's self-damaging energy insecurities and surrender to the US, NAFTA and foreign ownership that blocks rational energy policies. Laxer shows the road to energy security through domestically controlled transition fuels while top priority renewables and conservation rapidly displace a high-carbon economy. Read this book, Oh Canada, before the next energy shock.

—Ralph Nader, political activist, lawyer and author

This provocative and compelling book, written by one of our most insightful analysts, is an easy-to-read crash course in Canada's energy future. Gordon Laxer shows how, by breaking the control of the petro-elites, we can build ourselves a low-carbon future while ensuring we don't freeze in the dark in the coming energy crunch.

—Linda McQuaig, journalist and author of *It's the Crude, Dude: War, Big Oil and the Fight for the Planet*

Does it make sense for Canada to export more of its tar sands oil to the United States? Conventional economic wisdom says yes; Gordon Laxer says no, and argues his case with convincing evidence and logic. Why be North America's gas pump when Canada could be a green powerhouse? Laxer asks questions and highlights facts that may seem obvious but that somehow elude 'serious' pundits and politicians. Every Canadian should hear him out.

—Richard Heinberg, Senior Fellow, Post Carbon Institute and author of *Afterburn: Society Beyond Fossil Fuels*

The extraction of Alberta bitumen dooms both control of climate change and Canada's transition to clean energy. No one has made this linkage more persuasively than Gordon Laxer, in spite of politicians' repression of his right to express such sound professional and moral concerns. Many regard the Parkland Institute as one of the world's 'finest models for applied scholarship,' and this feisty book brings to a wider audience the vital research that has made Laxer so well respected—and so rightly feared by Canada's most-damaging polluters.

—Patrick Bond, University of KwaZulu-Natal Centre for Civil Society, Durban, South Africa and author of *Politics of Climate Justice*

To Judith, Damon, Christopher and Daniel—my guiding lights

In memory of Larry Pratt

Acknowledgements

It was late November 2004 in Calgary and it already felt like winter. Coming in from the cold, about 50 people settled into the Citizens Inquiry on Canada-US relations. The event was sponsored by the Council of Canadians and I was a commissioner with Maude Barlow and Ed Schreyer. The issue was energy.

An audience member asked why the US has an energy security plan and we don't. Another commented that although Brian Mulroney killed Pierre Trudeau's National Energy Program, we now have a new one—No Energy Plan. They got me thinking.

A year later, Parkland Institute brought 20 energy experts from across Canada to Calgary again, to help work out a Canadian energy and ecological security strategy. Thanks to those who shared their insights—Diana Gibson, Dave Thompson, Ricardo Acuña, Larry Hughes, Guy Caron, Bruce Campbell, John Dillon, Steven Shrybman, Mel Watkins, Hugh Mackenzie, Larry Pratt, Tony Pennikett, Chris Severson-Baker, Nasheena Sharif, Andrew Nikiforuk, Don McNeil and Keith Newman. The next month, Dave Thompson, with help from Diana Gibson and me, wrote a bold discussion paper for Parkland Institute. The ball got rolling.

I started an ambitious research plan and was greatly aided by four interns who far exceeded their research duties and challenged my ideas: Erin Krekoski, John Watson, Ryan Katz-Rosene and Tanya Whyte. Tanya continued to do superb research. Also, graduate research students at the University of Alberta's Sociology Department helped a lot: Ashok Kumbamu, Goze Dogu, Ineke Lock and Manoj Misra.

I benefitted greatly from many discussions with a loosely knit knowledge community—Andre Plourde, Andrew Nikiforuk, Dave Hughes, Dave Thompson, Diana Gibson, Erin Weir, Jeff Berg, James Laxer, John Dillon, Keith Newman, Kjel Oslund, Larry Hughes, Larry Pratt, Marjorie Griffin Cohen, Maude Barlow, Mel McMillan, Patrick Bond, Ricardo Acuña, Richard Heinberg, Rick Munroe, Thomas Homer-Dixon and Tony Clarke. I learned a lot from John Dillon about NAFTA's energy proportionality rule.

The contributions of those I interviewed were invaluable: Adam Ma'anit, Cyndee Todgham Cherniak, Daniel Breton, Doug Heath, Helmut Mach, Louis-Gilles Francoeur, Louise Vandelac, Lucie Sauvé, Matt Simmons, and Patrick Bonin.

I needed other eyes to check drafts for facts and interpretation and got much valuable feedback from Christopher Laxer, Duncan Cameron, John Dillon, John Ryan, Larry Hughes, Ricardo Acuña, Rick Munroe, Ryan Katz-Rosene, Shannon Stunden Bower and Tanya Whyte. I take responsibility for any remaining errors.

I owe a great deal to Bruce Westwood and Lien de Nil, my literary agents, who strongly supported the book (even though it was way too long) and went the extra mile in finding a great publisher.

Writing a book is a lot more than gaining knowledge and ideas. It must grab readers and sustain their interest. Good editing is crucial. The first version was too long and academic. Cathryn Atkinson cut overly detailed sections, shortened the manuscript and suggested *After the Sands* as the title. The editors at Douglas & McIntyre took it a giant step further: Anna Comfort O'Keeffe suggested major changes, Silas White moved and cut sections and put it all back together with the speed and skill of a magician, Merrie-Ellen Wilcox and Kyla Shauer's copy editing was thorough, helpful and meticulous, Teresa Karbashewski coordinated the excellent infographics and Petr Cizek made dramatic pipeline maps not seen before.

Books must be well written and designed but also widely available. Thanks to Bruce Campbell, Duncan Cameron, Kim Elliott, Maude Barlow, Meagan Perry, Ricardo Acuña and Trevor Harrison for support on circulation and grabbing the public's attention.

Thanks to my sons for their support. Christopher for great ongoing discussions and web design, Daniel for urging me to write a book rather than just articles and Damon for ideas on how to promote the book.

My wife Judith was my rock; she tolerated my many absences, listened to far too many of my formulations, questioned dodgy ones and urged me on.

—Gordon Laxer, July 22, 2015

Table of Contents

Preface   1

Chapter 1   "Let the Eastern Bastards Freeze in the Dark"   9

Chapter 2   Suddenly Without Oil   39

Chapter 3   Without a Parachute   71

Chapter 4   NAFTA and Proportionality: A Devil's Bargain   91

Chapter 5   Alberta: Fossil-Fuel Belt or Green Powerhouse?   109

Chapter 6   Resource Nationalism Everywhere but Canada   135

Chapter 7   Pipelines or Pipe Dreams   159

Chapter 8   Let Goods Be Homespun   185

Chapter 9   How Much Is Enough? A Conserver Society   203

Chapter 10 Solutions: Energy and Ecological Security
            for Canadians   221

Glossary   240

Endnotes   247

Index   272

Preface

Some say the world will end in fire,
Some say in ice.
From what I've tasted of desire
I hold with those who favor fire.
But if it had to perish twice,
I think I know enough of hate
To say that for destruction ice
Is also great
And would suffice.

—Robert Frost

L ethbridge, spring 2007. I was to give a talk on energy. As the founding director and head of Parkland Institute at the University of Alberta, I'd done this for years, speaking to smallish crowds, usually ignored by the media. But this time, two camera crews were already waiting, in the midst of a hubbub, when we pulled into the parking lot. Which celebrity were they looking for? As I stepped out of the car, a cry went up. "Please do that again—we want to get a proper shoot!" A bright, young TV reporter shoved a microphone in my face. "Why does Canada import so much oil when we have so much of it and export lots to the US?" she asked incredulously. "I didn't know that. Why would we do that?"

Why, indeed? It was a sensible question from someone too young to remember the bruising battles over the 1980 National Energy Program and the 1989 "free trade" deal with the United States. Television crews were lurking in that parking lot because a week earlier, Leon Benoit, the Conservative chair of the Standing Committee on International Trade, had halted my testimony as an invited expert witness in Ottawa. "Mr. Laxer is off-topic," he declared. The meeting was one of four to address the Security and Prosperity Partnership of North America (SPP).

The push for the partnership came right after New York's Twin Towers collapsed on September 11, 2001. Despite the North American Free Trade Agreement (NAFTA), transnational corporations straddling the Canada–US frontier found shipments stopped at the US border right after the September 11 attacks on New York City. Within weeks, the Canadian Council of Chief Executives (CCCE) and other interests who had brought Canada "free trade"

1

and NAFTA promoted further integration with the US. The CCCE speaks for the CEOs of the largest corporations in Canada. Many CCCE members, like Annette Verschuren, then head of Home Depot's Canadian branch, received marching orders from their US headquarters. The CCCE portrays Canada's national interest as coinciding with the interests of giant transnational corporations. The CCCE advised Canada to adopt the Bush regime's "security agenda," a sweeping plan that made America more authoritarian. In return, the CCCE hoped the US would unblock exports from Canada. Canada quickly signed on. The SPP included plans to gut environmental reviews that could delay approvals for cross-border carbon fuel projects.

The SPP's official goals were to enhance North American security and competitiveness. The real aims were different. Washington was building a wall around Fortress America. Giant corporations wanted Canada on the inside looking out rather than outside looking in. They wanted Canadian goods and services to enter the US as if they were domestic, something NAFTA was supposed to have already accomplished but hadn't. Parliamentary committee hearings were held in 2007, when hopes were still high that the SPP would win the day. The SPP's Canadian promoters wanted to keep citizens in the dark about threats to Canadian sovereignty. They feared that the more Canadians learned about the SPP, the less they would like it. But secrecy ultimately sank the SPP. Its cover was blown in several incidents, including the one I was involved in.

As director of Parkland Institute, I had headed a six-year study on globalism and its challengers, involving nineteen researchers in three countries. I agreed to testify as an expert witness because I was concerned that Ottawa's focus on ensuring US energy security would block Canada from addressing the triple crises of our time—climate change disasters, the end of easy oil and the need to secure adequate energy for all Canadians. The SPP was about the US getting even greater access to Canadian energy supplies. Exporting more oil would endanger Canadians' security of supply and greatly raise Canada's carbon emissions. That is why I went to Ottawa—to awaken the government, opposition parties and Canadians to the dangers.

In 2007, the minority government of Stephen Harper controlled the agenda in Parliament by making everything a confidence vote and daring opposition parties to defeat them and get blamed for causing an election. But the Conservatives didn't dominate parliamentary committees, and lost committee votes do not provoke elections. The Conservatives had only four of the ten members of Parliament on the Standing Committee on International

Trade. Peter Julian, a New Democrat MP from Vancouver, insisted that the committee hold public hearings on the SPP. MPs in other parties showed little interest until the committee needed unanimous approval to travel to Indonesia. Julian said he would approve only if the committee held hearings on the SPP. Four hearings were held in April and May 2007. Julian insisted that they be televised. The first three were. But no cameras were present on May 10, when I gave my testimony. Julian protested the cameras' absence as a ploy by Benoit, who dismissed the issue, blaming it on a last-minute venue change. The room was not set up for cameras. I was bemused by what I thought was partisan sparring. Cameras, though, would likely have inhibited what happened.

My trip from Edmonton was paid for by Parliament and ultimately by taxpayers. I felt an obligation to give the best possible presentation, limited in advance to eight minutes, and testimony afterward. When I began, I had no inkling that my appearance would shut down the hearings for good. My opening remarks caused a visible stir: "I don't understand why Canada is discussing helping to ensure American energy security when Canada has no energy policy, and no plans or enough pipelines, to get oil to Eastern Canadians during an international supply crisis...Canada is the most vulnerable member of the International Energy Agency, yet recklessly exports a higher and higher share of its oil and gas to the US. This locks Canada into a higher export share under NAFTA's proportionality clause. Instead of guaranteeing the US energy security, how about a Canadian SPP—a Secure Petroleum Plan for Canada?"

The Conservative MPs, who were seated in one corner, huddled in animated, whispered discussion, apparently no longer listening. Three minutes later, Benoit declared, "Mr. Laxer, I'm going to cut off your presentation." I objected, and this was noted in Hansard. Julian called the chair's position "absurd." *Ottawa Citizen* reporter Kelly Patterson described what followed: "Opposition MPs called for, and won, a vote to overrule Benoit's ruling. Benoit then threw down his pen, declaring, 'this meeting is adjourned,' and stormed out, followed by two of the panel's three other Conservative members."[1] The three Liberals, two Bloc Quebecois, one NDP member (Julian), and a lone Conservative (Ron Canaan) voted to finish the meeting.

Instead of dying in obscurity, the usual fate of a professor's testimony, mine was made intriguing by Benoit's actions. What did I say to provoke him so? Editors saw drama and protagonists. The next day, Canwest Global put Patterson's story in the first sections of the *Ottawa Citizen*, the

Montreal Gazette and the *Edmonton Journal*. The *Calgary Herald* and the *Edmonton Journal* published my testimony in full five days later. The following day I was in Lethbridge, surprised by the TV camera crews in the parking lot, and was handed an invitation to meet with the *Lethbridge Herald*'s editorial board. My opinion article on the incident soon appeared in the *Globe and Mail*. Few agreed publicly with Benoit that energy security for Canadians was off-topic in a discussion that covered "security" for North America. And that was it—my fifteen minutes of fame were over, and the media moved on. It was gratifying to see a groundswell of interest in Canadian energy security. The federal NDP and Greens have since occasionally raised Canadian energy security concerns, but silence from Conservatives and Liberals continues.

Nine months later, Benoit sent me the following gracious note after the Polaris and Parkland Institutes forwarded all MPs a copy of my research report: "I would like to thank you for sending me a copy of your recent report *Freezing in the Dark: Why Canada Needs Strategic Petroleum Reserves*. I found it to be an interesting [and] informative piece. I look forward to the day when you again appear before a committee I chair." An invitation never came. The SPP died a quiet, unheralded death in 2009, after Barack Obama became US president, but energy security for Canadians has not been addressed by Mr. Harper's government.

Despite its oil abundance, Canada is woefully unprepared for the next global oil supply crisis. The current oil price crash and temporary oil glut mask how vulnerable Canadians are. Beijing is cannily buying up to 700,000 barrels of oil a day to boost its emergency reserves, because it knows shortages will threaten people's economic and physical well-being in only a few years. Along with Australia, Canada is the most vulnerable member of the International Energy Agency (IEA). Canada imports 40 percent of its oil, yet—unlike twenty-six of the other twenty-eight IEA members—it has no strategic petroleum reserves to meet temporary shortages. Canada has no plan to meet the triple crises of climate change disasters, the end of easy oil and the post-9/11 shift to security trumping trade.

For forty-three years, Progressive Conservative governments ran Alberta hand-in-glove with Calgary's oil patch, giving them sweetheart deals on royalties and corporate taxes and turning a blind eye to enormous environmental damage. That's no longer assured. The "99 percent" finally challenged their hold on May 5, 2015, and elected an NDP government led by Rachel Notley. Big Oil may still be in the driver's seat, but it need not win every contest.

I don't use the phrases *oil sands* or *tar sands*. These terms were used interchangeably by the industry in Alberta until the mid-1990s, when the industry and Alberta's Conservative government rebranded them *oil sands* to improve their image. The correct term, *bitumen*, is technical-sounding and awkward. Whenever I submit an opinion article to Alberta's major newspapers and use a term other than *oil sands* and explain why, the editors ignore what I've written and invariably change it to *oil sands*. In Alberta, it's almost impossible to avoid being part of the effort to rebrand the Sands. To lower the debate's temperature, I use the neutral term *Sands* throughout the book.

Opposition to the drive to export Sands oil is growing from indigenous peoples who are "idle no more" and a broad array of non-indigenous people who are joining them. Opponents worry about pipelines leaking, oil fouling the beaches of Stanley Park, tankers running aground off the coast of British Columbia or in the St. Lawrence, and flimsily built tank-cars jumping railway tracks and exploding, like at Lac-Mégantic. Proponents and opponents of Alberta's Sands are at loggerheads about every pipeline, every new Sands project, every duck found dead in a tailings pond. The battles emit lots of heat but little light. Few are taking the long view. What can Canada be like after the Sands? Will there be a new beginning?

Although they hold enough recoverable oil to last a century or more, it is doubtful Alberta's Sands will operate that long. The environmental price is too high. The rest of the world is moving away from carbon fuels because of impending climate catastrophes. When the world moves on and stops importing Sands oil, where will Alberta and Canada be?

We can jump ahead of this fate by phasing out Sands production ourselves in an orderly fashion. It would be a giant step toward protecting habitats in Canada and the us and drastically reducing Canada's carbon emissions. We don't need another book with "The End of" in the title. *After the Sands* is an optimistic, uniquely Canadian roadmap to a low-carbon future. The best way to cut excessive carbon emissions is by phasing out Canada's carbon-fuel exporting role. Canadians can determine how much carbon energy we burn, but not how much others do.

It's easy to outline an eco-energy security strategy. The hard part is getting there. How do we overcome the power of vested interests and the export mindset that block Canadians from getting secure and fair access to Canada's own energy resources, enforcing tough environmental rules and introducing the polluter-pay principle? *After the Sands* presents ways to do this. If we phase out Canada's exports of conventional oil and natural gas, we will have enough to transition Canada to a low-carbon, conservation society. Existing hydro dams give us a large base of renewable energy to which we can add more environmentally acceptable renewables: wind, solar, geothermal and biomass. Few countries have these options in substantial quantities. We should make the most of them. Canada will not get there, though, until it shifts direction and removes Big Oil's hands from Canada's steering wheel. It can be done.

This is the first single-authored book to simultaneously explore the implications of climate change and the end of easy oil for Canada. The two issues are vitally connected. The best solution for one—a managed equitable reduction in carbon energy use—is also the best solution for the other. *After the Sands* takes a national view from a progressive Albertan and Canadian perspective. It is pro-Canada, pro-Alberta and pro-environment.

Most of us have multiple identities. Mine is Canadian, Albertan and Ontarian. I've lived for over three decades in each province. I was proud to get the Alberta Centennial Medal in 2005 for achievements that benefited fellow citizens, community and province. As director of Parkland Institute, I had a keen interest in creating a positive, eco-energy future for Canada. Unlike most think-tanks in Canada, Parkland Institute is not corporately funded; it does research on public policy for the common interest.

What Canada does matters to Canadians and the world. Action must be taken at all levels to lessen climate change chaos and protect habitats. Gaining international agreement is crucial because as long as a significant part of the world stands outside an international accord to cut greenhouse gases (GHGs), carbon-intensive industries will move to those areas to dump their GHGs into the air. Work at national levels is vital, though. It's where political power resides—the highest level at which citizens and democracy can effectively exercise their power or potential power.

Work at the local level is also imperative. Hopeful initiatives are springing up in many quarters, from Rob Hopkins' transition initiatives that have spread to more than a thousand communities around the world to food sovereignty campaigns led by La Via Campesina, to indigenous movements blocking oil pipelines and leading the way to humans living more harmoniously with nature. Local initiatives struggle to gain traction unless their national governments support them or at least do not block them. As *Canadian Dimension* publisher Cy Gonick argues, we must "defend the local nationally."[2]

Along with Australia, Canada is the developed world's most oil-insecure country and one of the world's worst polluters. Canadians will soon learn of their federal government's negligence the hard way, during the next major international oil supply crisis. It's during crises that most people pay attention and many open their minds to alternative paradigms. If left to business as usual, low- and middle-income Canadians in all provinces may shiver in the dark in future because they cannot afford to adequately heat or power their homes. As the world's easy carbon energy reserves run down, energy prices will skyrocket.

The best way to ensure security of supply is by getting domestic conventional oil and natural gas to all Canadians in a planned transition to a low-carbon society run on renewables. (Fracked and Sands oil are not conventional oil.) It's crucial to have thought-out alternatives beforehand so that when we search for answers, good ones will be available. That's why I wrote this book. The first seven chapters lay out the problems. The final three present solutions.

Chapter 1
"Let the Eastern Bastards Freeze in the Dark"

My father rode a camel. I drive a motor [car]. My son flies a jet plane.
His son will ride a camel.

—Sheik Rashid bin Saeed Al Maktoum, Saudi Arabia

In 1872, Jules Verne wrote a futuristic book about Phileas Fogg and his French companion Passepartout travelling around the world in eighty days to win a £20,000 wager. The world was much larger then; cheap oil and internal combustion engines have shrunk the earth since. While rapid travel is new, oil is not. Humans used oil thousands of years ago. The Chinese refined and moved oil through bamboo pipelines in 2000 BC. Egyptians lit lamps with crude oil in 1500 BC. Alexander the Great used flaming torches made with liquid petroleum in battle.[1]

Bitumen, a semi-solid form of petroleum found in great abundance in Alberta's Sands, has an even older history. Bitumen has been found on stone tools used by Neanderthals 40,000 years ago. Egyptians used bitumen to mummify their dead. *Mummy*, in fact, comes from the Arab word *mumiyyah*, meaning bitumen.[2] Bitumen, pitch, tar or asphalt—words used interchangeably—was long used for waterproofing ships, baskets and wood footings of bridges in rivers. The Bible talks about Noah using pitch or bitumen to seal the ark. The Egyptians waterproofed their ships with bitumen in 2500 BC. Fur traders in Alberta found indigenous people using bitumen to seal their canoes along the Athabasca, the great river that runs through the immense Sands region of northeast Alberta.

Oil wells were drilled in Russia, Poland and the Caspian Sea area of Azerbaijan in the early 1800s. The first successful North American well was punched in 1858 in Petrolia, Ontario, followed the next year by America's first near Titusville, Pennsylvania. Most oil was used to light lamps, replacing the hunted-out whale oil of *Moby Dick* fame. Petroleum oil became important only around 1900, when internal combustion engines began to replace horses, trains and coal-fired engines. Once oil became crucial in modern transport, the military wanted to store reserves. In the buildup to the First World War, oil-burning ships proved much faster than those burning coal. Winston Churchill, as Britain's First Lord of the Admiralty, switched the fleet from coal to oil in 1911 and stressed Britain's strategic need to procure oil. The next year, United States President William Taft set up an emergency naval oil reserve in Wyoming, called "Teapot Dome."

Oil got its land debut in September 1914 after France's government fled Paris in the face of advancing German forces. Taxi drivers proved oil's military value by rapidly rushing French troops to the front and saving Paris. Three years later, still at war, France ran short of oil and obliged oil companies to reserve ninety-one days' worth of oil for domestic use. Almost sixty years later, the International Energy Agency and the European Community set ninety days as the standard for strategic oil reserves.

Strategic petroleum reserves (SPRS) were also created by western governments during the oil supply disruptions of the Arab oil embargo in 1973–74. The SPRS were too late for that crisis, but are today's first line of defence against oil shortages. In the US, oil was first put into SPRS in salt caverns near the Gulf of Mexico in 1977. Fresh water injected into the caverns dissolved the salt and opened space to hold the oil. But the SPR failed its first test. After the Shah of Iran was overthrown in 1978, to avoid upsetting its key Middle East oil ally, Saudi Arabia, the US failed to release SPR oil leading to the biggest oil withdrawal ever from world markets: 5.6 million barrels per day.

Ironically, SPRS grew large enough for effective use only in the 1980s, when they were no longer needed. New liberalized markets combined with deep conservation made SPRS redundant. However, they proved useful later, during wars and natural disasters.

Canada Is Recklessly Unprepared

No one can be sure when the next international oil supply shock will strike. Despite oil's relatively low international price and the short-lived burst of US shale oil production, a disruptive international oil supply crisis will very likely hit in the next decade. Countries lacking at least one of the following conditions will be hit hardest:

- citizens having first access to domestic energy resources
- long-term oil supply contracts for oil-importing countries
- the military might to commandeer other countries' oil
- strategic petroleum reserves, or
- a transition already underway to deep-energy conservation and to renewables

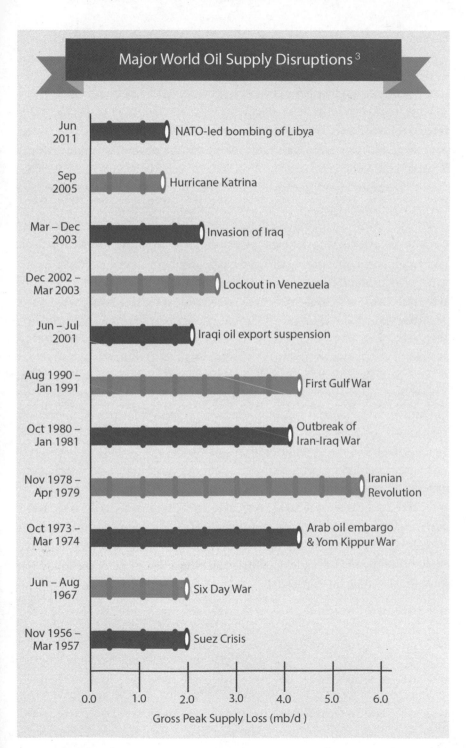

Major World Oil Supply Disruptions[3]

Jun 2011	NATO-led bombing of Libya
Sep 2005	Hurricane Katrina
Mar – Dec 2003	Invasion of Iraq
Dec 2002 – Mar 2003	Lockout in Venezuela
Jun – Jul 2001	Iraqi oil export suspension
Aug 1990 – Jan 1991	First Gulf War
Oct 1980 – Jan 1981	Outbreak of Iran-Iraq War
Nov 1978 – Apr 1979	Iranian Revolution
Oct 1973 – Mar 1974	Arab oil embargo & Yom Kippur War
Jun – Aug 1967	Six Day War
Nov 1956 – Mar 1957	Suez Crisis

Gross Peak Supply Loss (mb/d)

Despite vast oil deposits in Alberta's Sands, Canada lacks all five safeguards. It is urgent that Canada do what every other industrial country is doing: develop a *national* plan for global oil supply crises.

In 2012, Prime Minister Harper revealed why Canada has no Ottawa-initiated energy security plan, when the CBC's Peter Mansbridge asked, "Does it not seem odd that we're moving oil out of Western Canada to either US or new markets to Asia when a good chunk of Canada itself doesn't have domestic oil?"

"On a certain level it does seem odd," Harper replied. "The fundamental basis of our energy policy is market-driven...I think it served the country well; it served government revenues well; it served the creation of jobs well; but it is fundamentally a market-driven decision. We don't dictate pipelines go here or there."[4] In other words, Canada has no national energy plan because the prime minister doesn't believe in one. Harper added, "We're the only supplier that is secure." Secure for whom? Canada promises the US oil security, as the US has its own national energy security and independence plan. The question is, if Canada is looking after US oil security, and the US is looking after its own oil security, who is looking after Canada's?

The US presumes it has the right to help itself to as much Canadian oil as it wants, but remains a fierce energy nationalist. A 1975 act to put the US "solidly on the road to energy independence" bans the export of US crude oil, with only modest exceptions authorized by the president.[5] The 1920 Jones Act, still in force, requires all goods transported by water between US ports be carried in US-flagged ships built in the US, owned by US citizens and crewed by Americans.[6]

If Harper rejects a national energy plan, why the about-face by his provincial Conservative counterparts and Big Oil? Fresh from winning a mandate in April 2012, Alberta's Conservative premier Alison Redford convinced other western premiers to support a "Canadian energy strategy." She used buzzwords about energy efficiency, renewables, eco-regulations, cumulative impacts and "people as Alberta's most important resource," but made it clear that Alberta needs a national strategy because "we know we can't get our products to market without infrastructure that crosses other provinces."[7]

In 2015, Alberta's New Democratic Party (NDP) premier Rachel Notley is a less keen salesperson for exporting Sands oil. She's said she will not go to Washington to lobby for the Keystone XL pipeline, which she's declared to be an internal American decision. She also thinks the proposed Northern Gateway oil pipeline to the northern British Columbia coast creates too many environmental problems and too much opposition. But Notley's NDP government may represent more continuity than a break. Notley is open to the twinning of the Kinder Morgan oil pipeline to Vancouver and the Energy East oil pipeline to Saint John, New Brunswick. Those may be enough to allow continued expansion of Sands oil production.

One impetus for Alberta and Big Oil to support a Canadian energy strategy was to prevent a truly national strategy from taking hold. In 2005, Parkland Institute at the University of Alberta developed an Energy and Ecological Security vision for Canadians. As Parkland Institute's head, I helped drive this initiative, which shared the national energy "security" focus of official US and British plans but differed from them by challenging the supremacy of the petro-elites.[8] Petro-elites include transnational oil and gas corporations, both foreign- and Canadian-controlled, and the federal and provincial governments that support them. Regulatory bodies like the National Energy Board, set up to ensure that Canadian energy be developed in the national public interest, have been captured by the corporations they are supposed to oversee.

An Alberta Way Forward

Parkland's strategy took a unique view: a national eco-energy plan for Canadians from an Alberta perspective. Parkland was the first to adopt the term *Canadian Energy Strategy* to distinguish it from Pierre Trudeau's National Energy Program (NEP), the supposed dragon that mythically slayed Alberta's economy in the 1980s. (The NEP is continually denigrated in Alberta in the hope that a successor will never rise; it is never mentioned that the oil industry was devastated *around the world* during the time the NEP was in operation.) Because the Parkland strategy was critical of the Sands from inside Alberta, Big Oil and its supporters tried to head it off by usurping the "Canadian Energy Strategy" language and flipping it into a plan to support Big Oil's agenda. To succeed, the meaning of "national" had to be altered.

"National" energy security conjures up different imagery in Canada and the US. In the latter it usually means oil and natural gas self-sufficiency or energy independence, the kind celebrated on the Fourth of July. President Obama made it clear that Canadian oil is foreign oil when he declared that oil delivered via the proposed Keystone XL pipeline won't even stay in the US: "It is providing the ability of Canada to pump their oil, send it through our land down to the Gulf, where it will be sold everywhere else."[9] In Canada, "national" is employed to mean exporting unlimited amounts of Canadian energy that mainly enrich foreign energy corporations, regardless of environmental or energy security consequences for Canadians. Canada, alone among industrial countries, gave another country virtually first access to its energy resources through the North American Free Trade Agreement's (NAFTA) energy proportionality rule, an issue explored in chapter 4.

American environmentalists often pitch green energy proposals as enhancing US energy independence and security. Canadian environmentalists seldom talk about their proposals as helping energy sovereignty, because such a subversive idea would not be well received by the Canadian petro-elites. Our petro-elites promote *North American*, not Canadian, energy security because they want no limits placed on the sale of Canada's carbon fuels.

Canada can ensure energy security for all its residents, but only if Canada wins energy independence. Instead of leaving Eastern Canadians dependent on risky oil imports, we can ensure energy security for all Canadians by diverting domestic conventional oil (not Sands oil) currently bound for the US to Quebec and Atlantic Canada instead. Limiting production to Canadian *conventional*, non-fracked oil would threaten Alberta's Sands operators, but would be good in the long run for Albertans and all other Canadians. Restricting production of domestic carbon fuels to solely meet Canadian needs would mean drastic cuts to oil production and greenhouse gases. It would be good for the earth's climate but would slash corporate profits and asset values—the real reason that petro-elites fiercely resist Canadian energy and ecological security.

Along with Australia, Canada is the most vulnerable country in the International Energy Agency (IEA) to short-term oil supply shocks. Until 2014, Canada imported about 40 percent of the crude oil that our residents use, while exporting 70 percent of domestic crude oil output to the US.[10] The reversal of Enbridge's Line 9 from Sarnia, Ontario, to Montreal will cut crude oil imports somewhat, but the surge of US shale oil ensures that Enbridge's pipeline and railways will carry American as well as Western Canadian oil to Ontarians and Montrealers.

The US now supplies Canada with about half of its crude oil imports and also sends significant volumes of refined oil, including gasoline, to Eastern Canada.[11] Growth in American crude oil has displaced Algeria and other Organization of Petroleum Exporting Countries (OPEC) countries, which had supplied Canada with half its crude oil imports as recently as 2012. Relying on the US rather than OPEC countries for crude oil imports seems reassuring. But it should not be.

What will Washington do if an international oil supply crisis emanates from instability in the Middle East? Asked what would happen to US oil shipments to Ontario and Quebec in an international supply crisis, Matt Simmons replied, "It's pretty simple. We'd shut you off." Simmons headed one of the world's largest energy investment banks and was an energy advisor to George W. Bush.[12] Washington would stop US oil exports to anywhere else, too. The US Energy Information Administration forecasts that despite the surge in fracked shale oil, the US will import a quarter to a third of its oil through 2035—and national security always trumps anything else in the US. Canadians should be under no illusion that just because we send the US a lot of oil the US will reciprocate, or that NAFTA would prevent the US from ending the export of crude and refined oil to Canada. Washington has repeatedly ignored NAFTA on Canadian softwood lumber exports and would likely do so again on a commodity as strategic as oil.

Unlike twenty-six of the other twenty-seven International Energy Agency members, Canada has no national strategic petroleum reserves (SPRS). Nor does Ottawa have contingency plans to ship domestic oil to Eastern Canada, the only vulnerable part of the country, when a supply crisis hits (Western Canadians almost exclusively use their own oil). As federal Green Party leader Elizabeth May explains, "If there was a blockade of foreign oil or an economic embargo, those in Eastern Canada would have to wait for tankers to bring them bitumen for processing through the Panama Canal and up the eastern seaboard. As bizarre as that sounds, it was the solution offered by a Suncor executive when asked in committee about the vulnerability of Eastern Canada to embargos."[13]A shortage of furnace oil in Cape Breton, described in chapter 2, could be a harbinger of oil interruptions in Eastern Canada.

US Oil Surge Overblown

An impressive shale oil boom that started in 2009 reversed four decades of decline in US oil output. Many pundits have concluded that this boom heralds a new age of oil abundance, dismissing the idea that world oil production will peak any time soon. The boom is likely overblown and short-lived, though. By 2008, US crude oil production had dropped to 52 percent of its 1970 peak. By 2013, it had recovered to 77 percent and by 2019 it is projected to reach 9.6 million barrels per day, almost equalling its 1970 level. Then American oil production will shrink again, likely quite rapidly. By way of comparison, US oil consumption is 18.5 million barrels per day, double its peak domestic crude oil production forecast for 2019.[14]

Production of oil equivalents, including from natural gas liquids, will likely grow, too, but combined with domestic oil they will not meet US demand (further discussed in chapter 2). Geoscientist David Hughes shows that productivity in new oil plays in South Texas (Eagle Ford) and North Dakota (Bakken), which account for over 80 percent of US shale oil production, dropped by 60 percent after a year and to less than 40 percent by the second year.[15] In 2014, partly in response to Hughes's report, US officials adjusted their wildly rosy estimates of recoverable shale oil from the Monterey, California, area to 4 percent of their earlier projection.[16] The US will likely not free itself soon from dependence on foreign oil. To do so would require concerted action to reduce consumption to European levels. Swedes and Britons live as well as Americans do but on a third to less than half the amount of oil used per person. With 4.4 percent of the world's people, the US consumes 21 percent of global oil,[17] sucking in 7.5 million barrels of net foreign oil a day. Meanwhile, global conventional crude oil production peaked in 2006. With a few ups and downs, it has not budged notably since. New oil output has been balanced by about an equal rate of decline in existing oil wells.[18] More refined oil has recently been made from biofuels and natural gas liquids, but these sources aren't oil and cannot be sustained for long.

Canada's Climate Change Inaction

While Ottawa's Conservative government will not bring Canadians oil security, a Liberal federal government is unlikely to either. Liberal leader Justin Trudeau is eager to cast off his father Pierre's negative reputation in the West due to the 1980 National Energy Program. Thomas Mulcair's NDP might push for oil security, but it would have to reverse its drive for respectability with the elites and demonstrate a level of courage it has lacked so far.

Canada faces coming oil and natural gas shortages. Like every other country, it also faces climate disruptions. Evidence is overwhelming that human-generated greenhouse gases (GHGS) are causing climate change severe enough to threaten human lives and other species' survival. To prevent the world from heating up more than two degrees Celsius, climate scientists at London-based Carbon Tracker calculate that humans can emit no more than about 565 gigatons of carbon dioxide (CO_2) between 2011 and 2050.[19] Even a two-degree rise will cause great devastation. Anything more spells disaster unparalleled in recorded human history. The problem is that the world's coal, oil and natural gas corporations already hold 2,795 gigatons, or five times more carbon than the 565-gigaton limit.[20]

Contrary to Prime Minister Harper's claim in New York City in 2013 that

GHGS from Alberta's Sands are "almost nothing globally," [21] Canada's GHGS matter. With one two-hundredth of the earth's people, Canada produces one-fortieth of the earth's GHGS, as measured by production. That's five times the world's per capita level of emissions. Domestic oil and natural gas production is Canada's biggest source of GHGS, edging out transportation.[22] The Sands are also Canada's fastest-growing source of emissions and the main roadblock to meeting Canada's international climate commitments.

In the eyes of environmentalists and foreign governments, Canada is one of

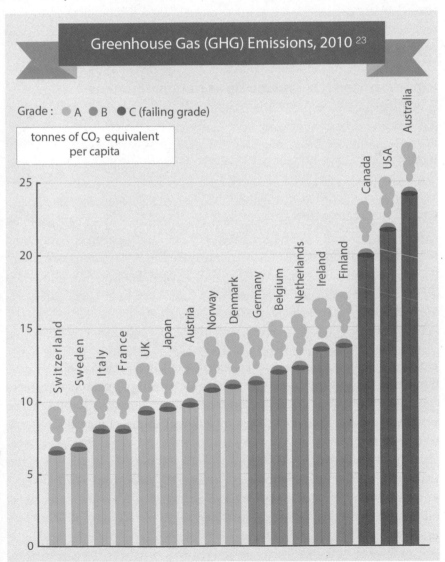

the worst eco-culprits, which is partly deserved and partly not. The deserved part is Canadians' high level of carbon use—sixth in the world per capita (almost a third less than Americans and 21 percent more than Britons).[24] The undeserved part is that all the carbon embedded in Canadian energy exports for consumption elsewhere, mainly the US, is attributed to Canada. If Canada phases out carbon energy exports, our country's emissions will plummet. Canada can and must cut GHGs by at least 80 percent from our 1990 level by 2050. That's the target Canada's House of Commons set in 2008, the first elected chamber in the world to approve legislation to cut emissions by that much.[25] The US and Europe soon endorsed the same target. It's a noble goal, one we must start to act on immediately in order to achieve.

Exporting Sands Oil Dressed Up as National Strategy

It was strange to see the words "national," "energy" and "strategy" strung together and promoted by Alberta's Conservative premiers. For thirty years, Pierre Trudeau's 1980 National Energy Program (NEP) was repeatedly trotted out, strung up and ritualistically flogged by Alberta's political and oil patch elite. Sentiment reached such a fever pitch in 1981 that Alberta stopped shipping 180,000 barrels of oil a day to other Canadians. Giant rallies in Calgary and Edmonton called for Alberta to separate from Canada. Ralph Klein, Alberta's Conservative premier from 1992 to 2006 made "get government out of the business of business" his focus. He later admitted that "we had no plan." Two decades after the NEP was dead he exclaimed, "By God, Ottawa, keep your hands off," even though no Liberal in Ottawa had threatened to put their hands on.

Energy plans were promoted, pilloried and changed elsewhere. A few adopted an energy conservation paradigm. But not Canada. After the traumatic battles over the NEP, energy policy was considered so toxic that no major political party touched it, thereby leaving the petro export model in place. Scientist David Suzuki described the impasse on Canadian energy policy this way: "The National Energy Program was implemented by the federal Liberal government in 1980 partly in response to skyrocketing oil prices. When the Conservatives came to power in 1984, they dismantled the divisive plan. Although the program did accomplish some of its goals, reducing foreign ownership of the oil industry as well as our dependence on oil, its most lasting legacy was to entrench a great divide between the oil-rich west and Eastern Canada. Since then, no one has dared to even mention the idea of an energy strategy for Canada. Canada is now one of the only developed nations without a coordinated energy plan." The post-NEP era was stuck in the neoliberal Reagan–Thatcher mantra, "Let the market decide." In Canada it meant sending the US as much of our raw resources, especially Alberta Sands oil, as possible, as fast as possible. Few dared to dissent.[26]

The Age of Easy Oil Will End

A pure market ethos was overtaken by the US reaction to the September 9, 2011, terrorist attacks. "Security trumps trade" captured the new ethos. Governments provide security; markets cannot. Change in the US-led ruling ideology coincided with the run-up of international oil prices to between double and seven times their 2002 level of $20 a barrel. However, in the fall of 2014 international oil prices had fallen to their lowest level since 2009. Don't expect these lower prices to last too long. We're in a pattern of undulating hills and valleys on a plateau of world oil production. This is what "peak oil" looks like.

Easy oil is no longer there for the picking. Why else would Big Oil still be in the Gulf of Mexico when BP received $85 billion in penalties after its giant Macondo well blew out in 2010, fouling the Gulf and killing eleven workers? Why else would Big Oil keep putting billions of dollars into environmentally sketchy projects like Alberta's Sands, fracking with 750 toxic chemicals, or drilling in the Arctic if conventional oil were still readily available?

The thirty-year deep freeze on Canadian oil policy officially thawed in the warm Kananaskis sunshine in July 2011. Alberta's Conservative government dropped a bombshell by calling for a "national energy strategy." Although it was not defined, other premiers endorsed it in principle. Ontario Premier Kathleen Wynne and Quebec Premier Philippe Couillard, both Liberals, later added environmental conditions to their support for TransCanada's Energy East pipeline to New Brunswick: Ontario will import hydro power from Quebec and join Quebec in a cap-and-trade plan to gradually reduce emissions. (The jury is out on whether cap and trade will significantly reduce emissions.) These are welcome revisions, but none of them have enough teeth to change Canadian energy strategy much, especially in comparison to limiting the export of Sands oil.

The challenge for Sands promoters is that Alberta is landlocked. Production cannot expand unless Sands oil either moves across other provinces or territories to seaports, or has Ottawa's help in overcoming US opposition to "dirty tar sands oil." Thus Alberta courted other provinces—and what better way than through something called a *national* energy strategy? Alberta's initiative had been promoted behind the scenes by oil executives and their allies since 2006. Patrick Daniel, CEO of the pipeline company Enbridge, was first. At Toronto's Empire Club, Daniel called for a "national energy strategy," to overcome the veto power over energy projects that every stakeholder currently wielded. With a proper vision, Daniel hoped, people would stop opposing energy projects.[27]

The next year, corporate-funded think-tanks ran with the issue. Roger Gibbins, then president of the Calgary-based Canada West Foundation, coordinated roundtables around "Getting it Right."[28] A big-picture thinker, Gibbins knew that to win over non-Albertans, a pro-oil-sands stance needed reframing. He explained that Alberta's Conservative governments needed "the protection, or cover, of a Canadian energy strategy…Alberta needs to embed the oil sands within a broader Canadian energy strategy…The oil sands must be seen as part, but only part, of a more comprehensive national strategy…[29] If the discussion of a Canadian energy strategy collapses into an oil sands strategy, Alberta loses the possibility of truly national support."[30]

Bruce Carson, nicknamed "The Mechanic," had been a senior policy advisor in the Prime Minister's Office under Harper, overseeing the most contentious files, including climate change and Alberta's Sands. Carson led the national energy strategy initiative. In 2008, he left the Prime Minister's Office to head the Canada School of Energy and the Environment at the University of Calgary, a centre that under his watch was mostly funded by his former employer—Harper's government. While there, Carson organized meetings of top oil industry insiders, including the "Banff Clean Energy Dialogue." Carson aimed to end the "dirty oil" image of Canada's oil sands industry: "We really felt, given the history of this public policy theme" that the energy strategy "really had to start from outside government."[31]

Carson never completed his work. Shortly before the national energy strategy announcement in Kananaskis by Alberta's Conservative government, Carson was felled by scandal involving three separate federal inquiries, including an RCMP investigation, over a conflict of interest. The sixty-six-year-old allegedly lobbied Ottawa on behalf of his twenty-two-year-old fiancé. His checkered past—imprisonment for fraud and disbarment by the Law Society of Upper Canada for defrauding clients—came to light. Carson disappeared from view. Despite the Carson setback, the national energy strategy was supported by the C.D. Howe Institute, associations of Big Oil and natural gas corporations, provincial hydro authorities, and the Canadian Council of Chief Executives (CCCE), representing 150 of the largest corporations in

Canada, many of them foreign-controlled. With the groundwork in place, it was safe for Alberta's Conservative premier to step up boldly and announce a national strategy.

Neither Canadian nor a Strategy

The concept touted by Alberta Conservative governments was a throwback to Canada's colonial past, with foreign-controlled businesses exporting Canada's raw resources and taking the lion's share of unearned profits from nature's bounty. It's neither Canadian nor a strategy. A real national strategy would see the federal and provincial governments secure sufficient energy for all Canadians within a framework of steadily reducing carbon emissions. The petro-elites' "strategy" advocates the opposite—unlimited production and export of carbon energy and continued oil imports.

Not all Big Oil promoters back a national strategy. Former head of the C.D. Howe Institute Jack Mintz calls it highly dangerous and likely doomed to fail because provinces have conflicting energy interests. Mintz is a fiscal, tax and public policy specialist at the University of Calgary and sits on Imperial Oil's board of directors. While getting all Canadians to feel that Sands development is valuable to them is good, Mintz argues, discussion of a national energy strategy suggests Canada's economy needs micromanaging, an idea he detests. Mintz worries that a federal government could combine its responsibility for the environment with a Canadian energy strategy to stop many projects. Advocating for a national energy strategy could do more harm than good, Mintz reasons: "We should drop the idea fast."[32] Harper is also skeptical. "The honest truth is I don't know precisely what [a national energy strategy] means," he has stated.[33]

The premises behind NAFTA and its predecessor, the 1989 Canada–US Free Trade Agreement (FTA), were as follows. Give the US priority access to Canadian oil and natural gas ahead of Canadians. In return, the US will refrain from limiting energy imports from Canada. This dream scenario for US petro-corporations is not surprising. It was almost certainly inserted into the FTA and later NAFTA at the insistence of Alberta's Conservative government and the US. NAFTA's proportionality rule obligates Canada to make available to the US the same share of its oil, natural gas and electricity as it has shared in the previous three years. Currently, it's about 9 percent of our electricity, over 50 percent of our natural gas and over 70 percent of our oil. Under proportionality, exports can rise or fall through "market" changes, in effect by decisions made by Big Oil, but Ottawa cannot reduce energy exports to protect the environment—or Canadians, as it did in the 1970s. The proportionality rule has never been invoked, but its existence may deter Canadian governments from ending carbon exports or protecting residents' energy security.[34]

In hundreds of "free trade" agreements signed over the past twenty-five years, no other industrial country has signed away to another country first access to its energy resources.[35] By making the export of energy virtually mandatory in a treaty with the world's greatest power, Big Oil and its Conservative allies assumed future Canadian governments would not alter it. "Critics say the problem with the [FTA] is that under its terms Canada can never impose another NEP on the country," stated Pat Carney, Canada's energy minister during the FTA talks. "The critics are right," she continued. "That was our objective...If the Americans promise not to block our energy exports, we promise in turn not to turn off the tap on energy supplies shipped under contract."[36]

And yet twenty years later, the US and Britain were singing a different tune. The US House of Representatives set up a committee on "Energy Independence and Global Warming" to focus on weaning America off Middle Eastern oil, reducing oil consumption, raising domestic output and cutting carbon emissions. In 2008, Barack Obama campaigned around "energy independence," reducing "dependence on foreign oil," increasing "national [energy] security" and "pulling the six-months trigger" to end NAFTA. Meanwhile, Gordon Brown, soon to be Britain's Labour prime minister, declared that "every nation today is concerned about energy security."[37] Except Canada—at least not at the level of the federal government.

Parkland Institute drew twenty progressive energy experts to the Kahanoff Centre in Calgary to work out a "Canadian energy security strategy" in October 2005. Parkland's report called for Canadians to get first access to their own energy, guarantee sufficient supplies for everyone in the dead of winter and transition the country to a low-carbon future.[38] Canada would regain national and democratic sovereignty over its natural heritage and resources, live more in harmony with nature, and take an international lead in cutting carbon emissions.

These initiatives challenged Big Oil's position in Canada. A national Energy and Ecological Security Plan could resonate well with many Canadians but would leave little room for huge windfall profits gained almost for free by accessing vast energy resources that Canadians and indigenous peoples hold in common. The best way for the likes of Enbridge, Gibbins and Carson to head it off was to appropriate the "national strategy" wording, strip out "security" for Canadians and redefine "national" to mean exporting energy. Whether Alberta's national energy strategy was motivated primarily by trying to co-opt Parkland's initiative or to get provinces outside Alberta to endorse Sands oil exports, the dialogue around it has been positive. After Alberta's Conservatives promoted a national energy strategy, it's difficult to criticize an eco-energy security version of it. It's legitimate in Alberta to debate national energy plans again.

Keystone XL Delay Jolts Petro-Elites

Canadian energy policy was set in stone for decades. The earth trembled a bit in 2007 when advocates for Canadian energy security, sovereignty and sustainability debated promoters of Big Oil, the Sands and climate change denial. But those debates were mainly confined to policy wonks, while Joe and Josée public were little engaged. The ground *truly* shifted in January 2012 when President Obama blocked TransCanada's Keystone XL, a pipeline proposal to take Sands oil to giant Texas refineries. Canada's petro-elites had assumed US approval was a shoo-in. XL's deferral changed everything. No longer could they presume that American authorities would grant their every wish. Blocking Keystone XL violated the US side of the understanding underlying NAFTA on energy: Canada gave the US unrestricted access to its vast energy resources in return for unlimited access to the US market. Obama's decision broke the promise of access and the spirit of NAFTA,[39] creating an opportunity for Canada to make an equivalent move: phase out Sands production and oil exports to the US, and redirect domestic conventional oil to Eastern Canadians.

Blocking Keystone XL was a watershed moment. Stephen Harper called it a "wake-up call...we've simply got to broaden our markets."[40] He added, "the very fact that a 'no' could even be said underscores to our country that we must diversify our energy export markets...We cannot be, as a country, in a situation where our one and, in many cases, only energy partner could say no to our energy products."[41] Even if Washington approves the new route for Keystone XL that was being reconsidered by Obama in 2015, Canada's petro-elites will never again complacently assume that the US will take every drop of Sands oil offered them and that America's oil addiction will always defeat its environmentalists.

Halting XL showed the vulnerability of special oil interests. Unless Sands oil can reach tide water, growth in Sands output will be choked off. That's the strategy of environmentalists on both sides of the border. Despite rail's rapid growth in moving oil, pipelines are and will remain the main vehicle: they have so much more capacity and are much cheaper, which matters especially when oil prices are low. No new pipelines means little Sands expansion, little growth in greenhouse gases (GHGS) and the preservation of many northern Alberta wilderness habitats.

Under the new uncertainty around the US market, Sands promoters looked seriously at other markets for the first time. China's fast-rising oil appetite is the great lure, predicted to take over half the growth in world oil

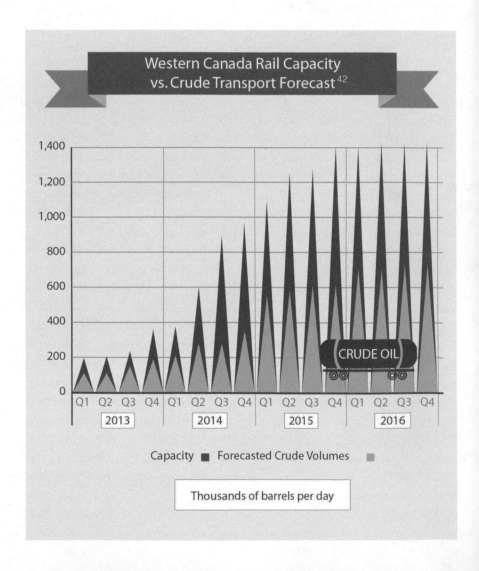

demand during the 2010s.[43] Two proposals— Enbridge's Northern Gateway to Kitimat, BC, and the twinning of Kinder Morgan's existing Trans Mountain pipeline to the Lower Mainland—will, if built, carry Sands oil to BC's coast for shipment mainly to China. Both pipelines face fierce opposition (further discussion in chapter 7).

The earth trembled under Big Oil's feet a second time when the Alberta New Democrats swept into power in 2015, astonishing themselves and everyone else. Before the NDP victory, Big Oil thought it had its Alberta bases covered. Its chief ally and Sands promoter, the Progressive Conservatives, were in government. If they fell, Big Oil had the Wildrose Party as backup. Seeded with lots of oil money, Wildrose surged from obscurity to become the Official Opposition in 2012, fighting against raising oil royalties in Alberta. The NDP victory ended the cronyism and absolute power of Big Oil in Alberta.

The implications of an NDP government in Edmonton for Canadian energy and environmental policies will become clearer over time. Notley has promised to switch funds for carbon capture to public transit. There will likely be tougher environmental regulations and a higher carbon tax on industry, and the provincial corporate tax rate will rise from 10 to 12 percent. Premier Notley is not at all beholden to Big Oil and will be a much less enthusiastic advocate for its causes. She is likely to join Canada-wide environmental plans and less likely to dig in around provincial rights.

But the NDP win in Alberta is not as big a political earthquake as many imagine, especially on energy and the environment. Notley's crew will likely be boldest on reducing inequalities and perhaps phasing out coal generation of electricity, but will likely tread most lightly around the special interests of oil and natural gas. The NDP's corporate tax increase is significant, but the rise will still leave it below that of the Conservative government of Ralph Klein before 2004. Notley may raise royalties, but if so they're bound to rest way below those of Norway, where Big Oil still thrives.

Because it intends to leave ownership of the oil and natural gas industry in private and foreign governments' hands, Rachel Notley's government will be more in office than in power. It will live in an uneasy cohabitation with Big Oil in the stronger position. Notley's government will depend heavily on the existing petro-owners for much of the private-sector job creation in Alberta and for revenue to fund its social programs. Worry about going too far and frightening petro-capital out of Alberta will be Big Oil's main way to tame the social democratic government in Edmonton. Will the NDP government and social movements take action to counter the threats? By taking significant action on the environment, and by obviously not being in Big Oil's pocket, the Notley government will likely soften outsiders' criticism of Alberta. Rachel Notley's premiership could come to be known for being a better promoter of the Sands than Big Oil's preferred political partners—or for being a leader in moving Canada to a low-carbon future. Time will tell.

Go East, Young Man

The push to diversify Sands oil markets beyond the US is not confined to BC. Enbridge is reversing the direction of its Montreal-to-Sarnia pipeline in order to take oil east to Montreal. It will move Western Canadian conventional oil, Sands oil and US Bakken shale oil. Pipeline rival TransCanada proposes to partly convert its natural gas mainline into an oil pipeline to Quebec and Saint John, New Brunswick.

No effort is spared to bring Sands oil to international markets. Billions in foregone profits depend on it, or so it is claimed. Former New Brunswick premier and former Canadian ambassador to Washington Frank McKenna pitched an oil pipeline from Alberta to Saint John a week after Obama halted XL. "We are painfully aware of the risks of putting all of our oil in the US basket," he wrote.[44] McKenna has a new career as a board member of Calgary-based Canadian Natural Resources (CNRL), one of the world's biggest independent oil and natural gas corporations.

McKenna's case is mainly economic. Staggering profits are lost because Canadian oil exports fetch much lower prices in the glutted US Midwest than they would from international sales, he contends. The price gap, labelled the "bitumen bubble," has allegedly cost Canada billions. Calls to end the bubble reached a fever pitch in 2013, as petro-elites fretted about delayed approvals for the XL and Northern Gateway pipelines. Trying to hide their self-interest, they contended that Canada suffered a "double discount" on the price it got for its oil. All Canadians were being hurt, they asserted. Enbridge's CEO Al Monaco claimed that delays on his corporation's Northern Gateway line are "a massive loss of value for Canadians."[45] Cenovus CEO Brian Ferguson called the lack of pipeline access a "subsidization to the United States consumer by the Canadian economy" of "$1,200 per Canadian." Alberta's Conservative government added a public interest angle by suggesting its deficit woes were a direct result of the growing bitumen bubble rather than its failure to adequately tax Albertans to pay for their public services.

The bitumen bubble claim is false. Bitumen has always gotten a lower price than West Texas Intermediate. Bitumen is priced lower because, as BC

economist and former CEO of the Insurance Corporation of British Columbia Robyn Allan states, "Bitumen is a junk crude that requires upgrading
and complex refining. It has always sold at a discount... Nor is it related to
pipeline capacity. The discount simply reflects the resource's poor quality." [46]
The price differential, however, is lessening. Ian Urquhart, political scientist
at the University of Alberta, has noted that Alberta's Royalty Review Panel
reported bitumen prices at about 55 percent of West Texas Intermediate
from 1997 to 2007; in the next four years, bitumen's price rose to roughly
80 percent. [47] The bitumen bubble was a made-up story to create support for
building Sands-exporting pipelines.

Robyn Allan has also shown that most Sands operators are not hurt by
a lower price for bitumen. Most are integrated corporations. What they lose
on bitumen's price, they recoup as refiners, with lower-priced inputs. Five
minutes after decrying bitumen's price differential, Cenovus's Ferguson
told an audience in Whistler, BC, "We are substantially benefiting [from the
wide differentials] at our refinery in Wood River." Furthermore, 30 percent
of Western Canadian crude already gets to "international markets along
TransMountain to the west coast or by way of Seaway and Pegasus to the
Gulf Coast." [48] If it doesn't get the world price, it's not because of lack of
pipelines, Allan argues.

Meanwhile, Frank McKenna waxed eloquent about nation-building as he
promoted the Irvings' interests, much like Alberta's Conservative premiers
wrapped the Canadian flag around a "national" strategy promoting Sands
exports. Comparing TransCanada's Energy East pipeline that would terminate at Irving Oil's giant refinery in Saint John to the building of the iconic
Canadian Pacific Railway, McKenna declared the coast-to-coast project would
be good for all regions, spreading oil-related jobs beyond Alberta. [49] What
McKenna *failed* to say is interesting. As a former New Brunswick premier,
you'd expect he'd promise New Brunswick oil security by replacing imports
with Canadian oil. But he couldn't do that because the Irving refinery wants
to mainly export. Nor does McKenna's nation-building vision include cutting carbon emissions, protecting habitats threatened by Sands projects or
recognizing Aboriginal land claims.

Derek Burney and Eddie Goldenberg added their voices to moving western oil to Asia via pipelines to Canada's east and west coasts. [50] Burney was
Brian Mulroney's chief of staff when the Canada–US Free Trade Agreement
was signed and the energy proportionality rule was first inserted. He sits on
TransCanada's board. Goldenberg was Jean Chrétien's chief of staff when

NAFTA began. Both urge Canada to quickly gain long-term oil contracts to supply China and India before supplying anyone else. Getting Alberta oil to Eastern Canada by existing pipelines would be faster than getting new pipelines to the west coast, they contend, even though shipping costs to Asia via the Panama Canal would be higher. Burney and Goldenberg portray corporate interests and Canada's interests as identical.

The Risk of Incidental Security

It could make commercial sense for TransCanada and Enbridge to deliver some domestic oil in Canada before loading the rest onto tankers for export, but even this incidental security would end when Big Oil finds it more profitable to export than to supply Canadians with our own oil. Energy security for Eastern Canadians would be reversed as easily as the direction of Enbridge Line 9 was in 1999. The forerunner to Enbridge Line 9, the Interprovincial Pipeline built in 1976, brought Western Canadian oil from Sarnia, Ontario, to Montreal for the first time, as part of Pierre Trudeau's energy self-sufficiency response to the 1973–74 oil supply crisis. The pipeline better insulated Canada from international oil shortages and skyrocketing oil prices by replacing imports with Canadian oil. Two decades later, for market reasons, Enbridge reversed Line 9's flow to bring foreign imports to Montreal again and then on to southern Ontario for the first time in decades. The federal Liberal government, gripped by the "magic of the marketplace" ideology, did nothing, and no major news story appeared on Line 9's reversal.

Current talk about sending Western Canadian oil east could repeat the earlier pattern. If international oil becomes cheaper than domestic oil, Enbridge or TransCanada could replace domestic oil with foreign oil and ship it west again. Also, west-to-east pipelines would force Eastern Canadians to buy Sands oil. Conventional crude would be mixed in the eastbound pipelines with more and more synthetic light crude made from bitumen. When buying gasoline, consumers do not have sovereignty. All gasoline stations, regardless of brand, get the same gasoline from the same refineries, whether it is derived from imports, domestic Sands oil or conventional oil sources, though domestic conventional oil won't be offered. How many Canadians want to aid in expanding the Sands eco-footprint?

The Case for Domestic Conventional Oil

Why do I insist that Canadians should use domestic *conventional* oil only? Oil from the Sands emits considerably more carbon and is destroying nature in much of northeastern Alberta, while tight or shale oil is fracked using hundreds of chemicals. Carbon emissions must be lowered so the atmosphere again holds no more than 350 parts per million. Carbon ominously passed 400 parts per million in 2013. In a low-carbon society, most energy would come from renewables, but conventional oil (and natural gas) will be needed for the transition. Unless Canada adopts a Canada-first eco-energy plan that steadily reduces our carbon emissions and protects local environments, west-to-east oil pipelines are just another way to peddle more Sands oil.

A report by BC energy researchers Marc Lee and Amanda Card shows that GHGs from Canada's gross exports of carbon fuels exceed emissions from all carbon fuels combusted in Canada.[51] Lee and Card show that Canada's massive export of carbon fuel prevents our country from meeting the Harper government's 2009 Copenhagen commitments to reduce GHG emissions 17 percent from 2005 levels by 2020—a reduction of 124 megatonnes. Canada cannot achieve this target if it keeps exporting so much carbon fuel. Sands emissions alone are forecast to triple between 2005 and 2020,[52] and most Sands oil is exported.

US Energy Security and Independence

The Security and Prosperity Partnership is dead, but powerful interests pressure Canada to continue to treat its energy resources as if they are "North American." Delighted that George W. Bush was gone, progressive Canadians' fears of losing Canadian sovereignty and being further integrated into the US lessened after Obama changed the face of America. But Democratic presidents such as Kennedy and Clinton have not recognized Canadian and Mexican sovereignty any more than Republican ones. Obama is not attuned to Canadian sovereignty issues either. Meanwhile, despite Harper's empty rhetoric about Canada as an emerging energy superpower, Canada is still committed to being America's gas tank and ensuring US, not Canadian, energy security.

South of the border, you can't avoid hearing weekly laments about US dependence on foreign oil and promises to end it. A 2011 Pew poll found that 91 percent of Americans agreed that US dependence on foreign oil was a very or somewhat serious threat, with 61 percent picking "very serious."[53] Invoking the spirit of 1776, Republican and Democratic politicians regularly promote US energy independence. Democrats have celebrated the Fourth of July as "energy independence day." Several Republican presidents have promised that the US will never again be held hostage to oil supply cut-offs.[54] (Especially on the Conservative side, Canadian politicians often copy their American counterparts, but not on *national* energy independence.)

In his 2006 State of the Union address, President Bush declared that "America is addicted to oil, which is often imported from unstable parts of the world." He pledged to "replace more than 75 percent of our oil imports from the Middle East by 2025." He did not, however, take steps to enable the country to recover from its addiction by bringing American oil use down to its national oil output level. Obama's promises went further than Bush's. Obama pledged to cut *all* oil imports from the Middle East by 2018. He added Venezuela to the cut-off list. American net oil dependence hit 60 percent in 2005, the year before Bush's "oil addiction" speech. Oil dependence is forecast to fall to 27 percent in 2015. Why the turnaround?[55]

US oil demand has been falling since the Great Recession began in 2008, while domestic crude oil production and liquid fuels, including gasoline from natural gas, are climbing.[56] Still, the US is not forecast to become petroleum self-sufficient. US profligacy makes it still hugely oil-insecure and pushes it toward trying to find oil under someone else's sands. Unlike Canada, the US has a National Energy Policy (NEP). Developed in 2001 and revised in 2007, it boldly proclaims national "energy security," "self-sufficiency," national ownership and control of petro-corporations.[57] Washington hounded Canada to ditch its NEP. It is ironic that the US now asserts similar economic nationalist themes for itself. Most people are in denial about the scale of changes needed to seriously tackle climate change threats and the depletion of easy oil. They hope new technologies will come along and make the problems go away. Every US president since Nixon has promoted complacency by promising technological fixes and US energy independence. Jimmy Carter (1977–1981) was astonishingly ahead of his time by being the only president to take conservation and the environmental damage caused by carbon fuels seriously.

In 1970, Richard Nixon began the "technology will save the day" tradition by declaring that he was "inaugurating a program...with the goal of producing an unconventionally powered, virtually pollution-free automobile within five years." George W. Bush painted fanciful pictures of new technologies and new energy sources, shimmering just over the horizon, and rebranded coal "clean." Comparing raising fuel standards to Kennedy's man-on-the-moon project, Obama's plan raises the efficiency of America's cars and trucks from

25 miles per gallon to 35.5 on model years 2012 to 2016. The fuel standards include trucks and suvs for the first time. The plan is supposed to cut 1.8 billion barrels of oil use over the lifetime of the vehicles sold. It's a far less inspiring goal than Obama's oratory suggests. In 2008, Europe, Japan and even China had already exceeded Obama's 2016 efficiency target. The program's end date conveniently falls at the end of Obama's second term.

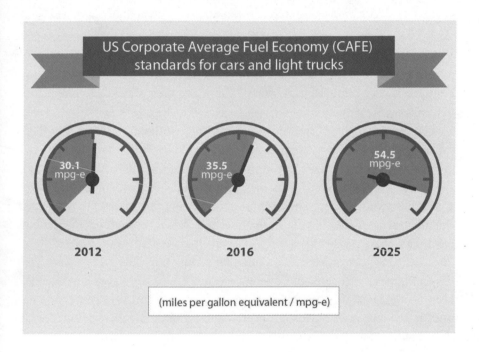

Obama later extended us efficiency standards to an impressive 54.5 miles per gallon by 2025. If future presidents hold to them, the standards will cut us oil use for cars and light trucks in half.[58] It would take until 2035 for the full effects to be felt. New vehicles stay on the road for at least ten years. Vehicle efficiency gains alone, though, will not end us oil imports. Cars and light trucks make up only 45 percent of total us oil use.[59] Obama's conservation plan is a start, but current us leaders fail to tell the public it will not simply be a matter of switching to new green-energy sources and electric cars and carrying on as wastefully and intensively as before. Americans will have to drastically reduce energy use. Dishonesty about the coming end of the era of easy oil is a failing in Canada, too.

Natural Gas and Canada's Recklessness

Natural gas is different. The temporary boost from shale, fracking and horizontal drilling pushed US natural gas production higher than it's ever been. The US is forecast to become a net natural gas exporter by 2016.[60] With one-fifth of world production, the US could easily end imports and become self-sufficient. It will likely import some natural gas, though, even as it ramps up exports.

Fracked oil and Alberta bitumen are non-conventional oil. Hydrofracturing (fracking for short) involves drilling into shale formations and injecting water, sand, ceramic and about 750 chemicals, such as hydrochloric acid and peroxodisulfates, under high pressure. It prevents fractures that open to release gas in tight formations from closing again when injections stop. Fracking is very contentious. Industry sees it as a great way to unlock oil and natural gas that would not otherwise be available. People who live in fracked areas often see it as an environmental disaster. Companies frequently hide the toxic chemicals they use in fracking.

Scientist Robert Howarth and his colleagues at Cornell University found that shale gas's GHG footprint is between 20 and 100 percent greater than coal's over twenty years. Their report contradicts the conventional wisdom that natural gas is the cleanest carbon fuel and should replace coal to generate electricity. Most US natural gas imports come from Canada. But they're falling and will continue to decline as Canadian production falls and US shale gas displaces them. The pace of Canada's natural gas exports is unsustainable, even reckless. Canada cannot long remain the world's fifth-largest producer and fourth-largest exporter of natural gas. Canada has only 1 percent of global natural gas—ranking twenty-first in the world. Canadian output has fallen by a quarter since peaking in 2006.[61]

Canada will have some domestic gas well into the future. Coal-bed methane, a far worse polluter than conventional natural gas, will offset some of Canada's shortfall. But adding coal-bed methane to new conventional gas finds will still likely result in decline.[62] Shale gas in BC's northeast region holds much promise, as do deeper, more productive conventional, tight and shale wells. It's not a matter of having no Canadian natural gas in future, but of not having enough, and not at a price low- and middle-income people can afford.

If Canada rapidly depletes its domestic natural-gas stock through liquefied natural gas (LNG) exports and wastes its supply on heating Alberta's bitumen, Canadians will likely depend on gas imports from sketchy suppliers like Russia and Algeria to stay warm in winter in the 2020s.[63] Canada squandered 1.5 billion cubic feet of natural gas a day in 2012 to make dirty Sands oil, the majority of which was exported to the US. Bill Powers, a Calgary-based energy analyst, forecasts the Sands will burn up 2 billion cubic feet a day in 2015, using almost 20 percent of Canada's total natural gas output.[64] Jim Dinning, Alberta's Treasurer under Conservative premier Ralph Klein, likened using natural gas in the Sands to reverse alchemy: "Injecting natural gas into the oil sands to produce oil is like turning gold into lead."[65] Americans don't need Canada's natural gas. We do. The sensible thing for Canada is to quickly phase out exports, seriously cut domestic use and phase out Alberta's Sands. If we do these things, Canadians can have natural gas supplies for decades as we transition off carbon fuels.

US Oil under Someone Else's Sands

Jimmy Carter was the only US president in the past fifty years prepared to make his country oil-independent in the only serious way—through major conservation so the US could live off its own domestic supply. Between Carter's green offensive in 1977 and Ronald Reagan's freezing of fuel efficiency standards in 1986, the US economy grew by 27 percent, yet oil demand fell by a sixth.[66] Economies can grow while oil use falls. If subsequent presidents had held Carter's course, the US would be energy-independent and the globe's environmental leader today. The Gulf wars to secure oil might not have happened. We look at Carter's plan in more depth in chapter 2. Ironically, Carter's plan, similar action in Europe and Japan, and the demand-killing effects of oil price spikes were too successful. Oil use fell and cheap oil returned for two decades. Conservation and new oil finds that high oil prices made economically feasible undercut OPFC's power. Reagan revelled in cheap oil's return and removed Carter's solar panels from the White House roof. Rising oil imports and greater oil insecurity resulted. Colossal US energy waste was back.

So was the return to finding "American oil" under someone else's sands. Jeff Gotbaum, a Bill Clinton regime official, told the Senate Foreign Relations Committee that the 1991 "liberation of Kuwait cost $57 billion or twenty-five cents a barrel of oil." Dan Plesch, a British security expert, considers that figure ludicrously low. The 2003 invasion of Iraq, the second Gulf War, was even more extravagant: "America's annual spending on the military and intelligence was around $400 billion and $40 billion respectively in 2004–5. About a quarter of this—or $120 billion—is focused on securing Middle East oil supplies." When America's allies costs are included, the tariff on trying to secure Middle East oil was over $20 a barrel, Plesch calculates. It was a failed strategy. Iraq's oil output fell by half after the invasion and did not return to its year 2000 level until 2011.[67] Imperialism fosters US oil dependence.

US Debate, Canadian Election

Washington counts on Sands oil to help wean the US off OPEC oil. But Sands oil cannot be produced with as low a carbon footprint as conventional oil. A powerful US movement to block "dirty" Alberta oil is growing. California's Low Carbon Fuel Standard, begun in 2007, will require oil companies to incrementally reduce the carbon footprint of all the steps in the fuel's production and usage by 2020.[68] In 2007 as well, the US Energy Independence and Security Act forbade federal agencies, including the military and the post office, from buying synthetic fuel whose life-cycle GHG emissions in making and combusting the fuel is more than those of conventional fuel.[69] The act clearly referred to Sands oil. Big Oil and its Congressional supporters have tried repeatedly to repeal it. Washington hasn't enforced it.

In an interview with the CBC's Peter Mansbridge, Obama brushed off concerns about massive carbon emissions from the Sands with the dubious assertion that carbon sequestration can solve their problems.[70] "Oil sands creates a big carbon footprint. So the dilemma that Canada faces, the United States faces and China and the entire world faces, is how do we obtain the energy that we need to grow our economies in a way that is not rapidly accelerating climate change."[71] American officials usually frame reducing "dependence on foreign oil" in national terms. Nevertheless, when Washington pledges to get off "foreign" oil, it doesn't mean getting off Canadian or Mexican oil.

What will Washington do if Canadians elect a federal government that phases out oil exports to the US and redirects the oil to Eastern Canadians? Canada did that after the 1973–74 oil crisis. Tough environmental action on the Sands is inconceivable under Harper. If a minority NDP or Liberal government has the fortitude to strictly enforce environmental laws on the Sands and the oil companies cannot comply, resulting in the Sands being phased out, could the US get by? Canada supplies 28 percent of US oil imports, almost equal to the 29 percent from Persian Gulf states and Venezuela combined.[72]

Phase Out Exports and Sands Oil

Phasing out the Sands and reserving domestic conventional oil for Canadians' use means ending Canadian oil exports to the US. It would be more difficult for the US to end oil imports from the Middle East and Venezuela. But because of timing, phasing out the Sands may not undermine the US plan much. Capping and then phasing out the Sands will take fifteen to twenty years, enough time for rising US fuel efficiency regulations to cut oil use substantially. Washington would have time to find new oil import sources. As a global power, the US has a security strategy of using many import sources, not relying on a single seller.[73] If Canada sends Eastern Canadians conventional domestic oil, it would free up several hundred thousand barrels a day that Canada imports. The US could import that oil or its equivalent.

Once the first major international oil-supply crisis hits, Canada will have no choice but to provide for its own residents first. When author William Marsden asked the late Peter Lougheed, former Conservative premier of Alberta and a prime booster of the 1989 Canada–US Free Trade Agreement, whether Canada would stick to NAFTA's mandatory, proportional, energy-exporting rule if Canadians ran short of oil, Lougheed replied, "If for some unusual reason we have a problem with supply, I think what would happen is the Canadian parliament, including support by the government of Alberta, would say, 'We've got to serve the Canadians first.'"[74] Lougheed thought the possibility remote.

Although US policies are inadequate to deal with energy security and sovereignty, climate disruptions and the end of non-fracked conventional oil, at least Americans debate and link those issues. Not Canadians. Many harbour frontier myths that Canada has so much land and resources that it can with impunity dump massive toxic wastes and tear off the boreal forest to get at Sands oil. Thinking they won't run out soon, we export the majority of our conventional oil and natural gas. Many Canadians support Sands expansion although they know it is dirty, because they've been convinced it's the best way to create jobs. According to a 2014 Environics poll, 41 percent of Canadians think the Sands contribute six to twenty-four times more economically than they actually do.[75]

Canada ignores the insecurity of importing so much oil and assumes the imports will always be there. These assumptions are wrong. Canada exports oil because the influential, foreign-dominated petro-elites know that exports mean greater profits. Canadians believe the US needs our energy, when with determination the US could become energy-independent. But Canadians need Canadian energy. Americans don't. To make the transition to a low-carbon society, Canadians must first change what's in our heads. Then we can confront the petro-elites that block the way.

Plenty of Public Ownership, None Canadian

Everyone is nationalizing their oil industry. Even former US Vice-President Dick Cheney acknowledged it. "Governments and the national oil companies are obviously controlling about ninety percent of the assets. Oil remains fundamentally a government business," he mourned. Most of the world's oil is off-limits to foreign and private investments.[76] Canada is one of the odd ones out. With the exception of Newfoundland, governments in Canada are not getting into owning and controlling oil and natural gas companies for the benefit of their people. Canada is one of the few oil-rich places left that is open to foreign ownership and control. Jeff Rubin, former chief economist at CIBC, calculates that Alberta's Sands represent 50 to 70 percent of all the private investable oil reserves in the world.[77] That's why foreign, state-oil corporations flock to Alberta.

There's plenty of public ownership in the Sands. "The only problem," notes Diana Gibson, former Research Director of Parkland Institute, "is that none of it is Canadian." She explained that citizens of Norway, China, South Korea, Japan, Abu Dhabi, Thailand and India profit as owners of Alberta's oil. "Public ownership is the best way to capture royalties, as 100% [would go] to the owners, the people of Alberta. It is also the best mechanism for ensuring appropriate development of the resource."[78] But oil security for their country, not profits, is the main reason so many state-owned oil corporations invest in the Sands. State-ownership stakes in the Sands have included some of the biggest state-owned corporations:

- China's National Offshore Oil Corporation (CNOOC) bought Calgary-based Nexen in 2012, giving it a major stake in the Sands. CNOOC also has holdings in Sands producer MEG Energy Corp.

- In 2012, PetroChina, also state-owned, expressed interest in buying into Enbridge's proposed Northern Gateway pipeline, which would likely ship much Sands oil to China. PetroChina bought Athabasca Oil Sands Corporation in 2009.

- Sinopec, another Chinese state oil company, joined with Synenco to form the Northern Lights project north of Fort McMurray and planned an upgrader.

- Statoil, two-thirds owned by Norway's government, bought North American Oil Sands Corporation. Thailand's state-owned PTTEP bought into StatOil's Kai Kos Dehseh, another Sands project.

- State-owned Korea National Oil Corp (KNOC) bought Newmont Mining's Black Gold project in 2006 and Harvest Energy Trust in 2009. Both have Sands holdings. KNOC aims to get Sands oil for energy-insecure South Korea.
- Abu Dhabi's TAQA, a majority state-owned company, bought three Sands companies in 2008.[79]

The Harper government limited national oil company takeovers of petro-corporations after China's CNOOC bought Nexen and Malaysia's Petronas took over Progress. Ottawa will bar such takeovers in future to "safeguard Canadian interests," allowing only minority control.[80] That's good, but why limit safeguards to foreign state-owned interests? Equally concerning is so much ownership and control by privately owned foreign transnationals, such as ExxonMobil and Shell.[81] They form a formidable, internal petro power bloc with local politicians that unduly influences Canadian policies over what domestic resources to exploit, their toxic ecological consequences, and who benefits from setting royalties and taxes on unearned, public natural wealth.

Concern about foreign control is a prime reason why so many Canadians—51 percent in a Leger poll in 2005—favoured public ownership of Canada's oil and natural gas companies.[82] If governments in Canada were responsive to their citizens, they would set a new legal framework to ensure that oil and natural gas companies serve Canada's public interest. Non-profit, enviro-energy companies could be incorporated under rules mandating them to wean consumers off carbon fuels, coax them onto renewables, and secure sufficient energy supplies for everyone.

Fanciful, False Futures

Governments delude citizens with false futures. It is a mirage that Canadians have unlimited oil and natural gas supplies and can afford to export the majority of them and not replace current oil imports with domestic conventional oil. The Sands will not save the day. Unless a miracle happens and their carbon footprint and horrific impacts in northeastern Alberta, a quarter of the province, are reduced to those of conventional oil, pressure to stop Sands expansion will grow.

It is ironic that Washington rather than Ottawa promises oil independence. If realism ruled, it would be the reverse. Given the Americans' record depletion of their once incredibly bountiful energy supplies, they cannot have cheap oil, very low gasoline taxes and a ludicrously wasteful energy lifestyle for much longer and gain national energy independence. On the other hand, Canada has the resources but needs the political will to become energy-independent, provide a secure minimum of energy to all its residents and become a low-carbon society run on renewables. Canadians should quickly join the paradigm shift that is sweeping across Western Europe. The end of abundant carbon fuels is near. Canadians must alter lifestyles built around cars and challenge the power of foreign state-owned and privately owned petro-transnationals. The longer we delay, the more catastrophic the fall will be.

With half of 1 percent of the world's people, Canada emits over 2 percent of its GHGs. When Canada signed the Kyoto Accord in 1996, it was committed to cutting GHG emissions to 555 million tonnes, 6 percent below their 1990 level, by 2012. Instead, Canada spewed out 699 million tonnes, 26 percent above that level, in 2012. Sands emissions are responsible for 36 percent of the growth of Canada's GHGs since 1990. In 2011, they emitted 55 million tons, the equivalent of nine million cars. Although the Sands accounted for only 8 percent of Canada's GHGs, their emissions are expected to rise sharply as Sands' output grows and other uses fall.[83]

When you destroy swaths of Alberta's boreal forest to get to the bitumen, a great deal of carbon is released, there is no forest to act as a carbon sink, and a lot of natural gas is burned making synthetic crude. Canada cannot greatly cut carbon emissions while the Sands keep expanding. Oil production is the fastest-growing source of Canada's GHG emissions, but 70 percent of domestic oil is produced for export. This means that when Canadians use way less oil, Canada's GHGs will not fall as a result. NAFTA's virtually mandatory exporting rule severed the umbilical cord between oil output and oil use in Canada.

Chapter 2
Suddenly Without Oil

Let the eastern bastards freeze in the dark.

—Popular bumper sticker in Alberta in the early 1980s

Unfortunately, the NEB has not undertaken any studies on security of supply.

—National Energy Board communications team's reply to author's query, April 12, 2007

Canada used to be better prepared for a prolonged international oil scarcity. In 1970, just before the two oil shocks of that decade, oil companies in Canada were 91 percent foreign-owned-and-controlled, mostly by Americans. Those oil corporations designed a continental, American distribution system that disadvantaged Canadians, much like today. Canada exported half (53 percent) of its oil to the United States, all from Western Canada, and imported half (49 percent) of the oil used by Canadians. All imports came in the east. Canada now exports 73 percent of its oil productions, while imports are down a bit to 40 percent and will likely fall further.

In 1970 no one talked about energy security for Canadians. Like today, the feds had no plans for it. There was little need for a natural gas security plan. Natural gas ran through a pipeline on an all-Canadian route from Western Canada to Quebec City, supplying six of the ten provinces, which accounted for almost 90 percent of Canadians. Domestic natural gas did not reach easternmost Quebec or any of Atlantic Canada. Oil was different. An oil pipeline from Western Canada reached Ontario after running through Great Lakes states, where it offloaded more than half its supply. The pipeline re-entered Canada at Sarnia, near Windsor. John Diefenbaker's 1961 National Oil Policy drew a line at the Ottawa River, reserving Ontario and the West for Western Canadian oil and preventing Western Canadian oil from reaching Quebecers and Atlantic Canadians. The five easternmost provinces relied entirely on oil imports. Oil had not yet been found off Newfoundland.

Following the 1973–74 Arab oil boycott of the us and the Netherlands, Canada quickly reshaped its energy policies. Pierre Trudeau's government adopted an energy "self-reliance" strategy and created Petro-Canada, a government-owned oil company that bought out one oil transnational after another. This coincided with a nationalization wave between 1970 and 1975 that de-globalized 336 transnationals in the world, many of them oil companies.

"Self-reliance" prioritized domestic oil for Canadians. An oil pipeline (Interprovincial Pipeline) was extended to Montreal (the forerunner of today's Enbridge Line 9), busting through the "Ottawa River line" and bringing Western Canadian oil to Quebec. To free up oil to send to Quebec, Canada reduced exports to the us and told Americans it would send them oil only if we had long-term surpluses. When the international oil price shot up over 1,000 percent between 1973 and 1979, Canada held down domestic gasoline prices but exported oil to the us at the world price. By 1981, Canadian oil exports had dwindled to 14 percent of their 1973 level.

The 1980 National Energy Program (NEP) was a successor to the self-reliance policy. The NEP aimed to raise Canadian ownership to 50 percent of the petroleum industry through government and private Canadian ownership. It was wildly popular, even in Alberta—at least at first. The NEP soon died, however. The world oil price collapsed in 1982. Although all oil-producing regions in the world were devastated, most Albertans blamed the NEP for Alberta's economic woes. The stage was set to overturn the NEP and use NAFTA's unique proportionality rule to lock in the change. There was a new NEP— *No Energy Policy*. Governments and political parties still shy away from taking effective action on energy and ecological security because of their fear of replaying the bruising regional battles of the early 1980s. After more than thirty years, it's time to get over it. To ensure energy security for Canadians and join the international battle against climate chaos, we must move on.

The Making of a US Energy Satellite

For eons, schoolchildren have heard that Canada evolved peacefully from "colony to nation." Harold Innis disagreed. "Canadian nationalism was systematically exploited by American capital. Canada moved from colony to nation to colony."[1] Canada enjoyed independence briefly between the world wars, Innis observed, as it shifted from "British imperialism to American imperialism." After a brief period of energy independence around the Energy Self-Sufficiency and the National Energy programs in the 1970s and early 1980s, with regard to energy Canada reverted to colony again with the FTA and NAFTA's proportionality rule. By the 1950s, America's natural bounty, particularly oil, was being depleted. In response, Washington and Big Oil cast covetous eyes on Canadian energy as the most secure source for their insatiable appetites.[2] Two roads to privileged access to Canadians energy were available: 1) US corporate ownership and control, and 2) reshaping Canadian energy policies to give priority to exports to the US.

First Road to Energy Satellite

US corporations began taking over Canadian oil when Rockefeller's Standard Oil bought Canadian-owned Imperial Oil in 1899. Foreign ownership and control of Canadian oil and natural gas reached an astonishing 90 percent by 1960.[3] Working from their dominant position within Canada, US petro-subsidiaries pressed Ottawa to approve exporting Canadian oil to the US. It made profit sense to their head offices to integrate Canada into their continental corporate systems. US corporate control over Canadian energy resources was part of a broader pattern of US corporations buying and using oil abroad to conserve domestic energy sources.

Although the US inherited perhaps the greatest store of oil in any country in the world, its full supply was unknown in 1940. Projections put reserves at only fourteen years of domestic supply. Just before Japan attacked Pearl Harbor, US foreign serviceman William Ferris outlined US oil options for the war everyone saw the US joining. Ferris and other State Department planners anticipated much higher oil use once the US joined the war.[4] The US had enough domestic oil for its own needs and for those of its wartime allies. But if it used only US oil, domestic supplies would be insufficient in future crises.

Ferris recommended that the US extend its domestic oil supply by using more foreign oil. Foreign oil dependence would be counterbalanced by holding US oil for later use. But other powers also sought oil from the same sources—the Middle East, Venezuela and Mexico. Ferris's solution was for US foreign oil policy to be "more and more aggressive" to ensure access to foreign oil.[5] Washington followed this road, and Ferris's solution to future American oil shortages was followed for most of the time between 1940 and the 9/11 terrorist attacks on New York City in 2001.

US president Franklin Roosevelt and Saudi King Ibn Saud made a pact in 1945 that still holds: the US will protect ·Ibn Saud militarily in return for privileged access to Saudi's huge cheap-oil reserves.[6] Ferris's boss, Max Thornburg, saw the main challenge as preventing the nationalization of US oil corporations, which Mexico had done in 1938. He encouraged Washington to use US companies as instruments of its foreign oil policy and protect them from nationalization.[7]

Jimmy Carter (1977–81) daringly steered a new US energy path. On television, he outlined a plan to make the US energy-independent the only realistic way—through conservation.[8] With the exception of war, Carter said, the 1970s energy crisis was the greatest challenge to America. He warned that "The oil and natural gas we rely on for 75 percent of our energy are running

out...Our nation's independence...is becoming increasingly constrained."
Carter laid blame where it belonged: "Ours is the most wasteful nation on
earth. We waste more energy than we import. With about the same standard
of living, we use twice as much energy per person as do other countries like
Germany, Japan and Sweden." If the US does not change, he said, "demand
will overtake production."

Carter tied energy security to environmental protection. "Our energy
problems have the same cause as our environmental problems," he stated,
"wasteful use of resources...Conservation helps us solve both at once."
Prophetically, Carter warned that his policies would not be easy or popular,
and would spark opposition from "special interest groups," a reference to
Big Oil. Pierre Trudeau (1968–79, 1980–84), Canada's prime minister during
most of Carter's presidency, was as bold in boosting his country's energy-
independence. But Canada partly clung to US coattails on conservation. As
the only major industrial country without a domestically owned auto industry, Canada was a passenger in the US drive to boost auto fuel efficiency.
Canadians bought fuel-efficient cars from US and non-American automakers
that adhered to the US rules. Canada had its own oil conservation and off-oil
plan, too.

If later presidents had maintained Carter's course, reducing use at the
rate of falling US oil and gas production, the US would be energy-secure,
energy-independent and the world's environmental leader. The two Gulf
wars to secure foreign oil would likely not have occurred. Building on the
previous presidents' conservation and security framework, Carter acted
on his words. Corporate average fuel efficiency (CAFE) standards came into
force under Carter. CAFE required auto companies to annually improve
average gas mileage. By 1985, US cars averaged 25 miles per gallon, up
from 15 miles per gallon in 1975.[9] Unfortunately, CAFE standards
excluded SUVs, minivans and pickup trucks, vehicles that soon exploded
in popularity.

Sadly, Carter reverted to the imperial route of grabbing other people's
oil to maintain American energy independence. When tensions with Iran
escalated after its 1978–79 revolution, Carter issued the Carter Doctrine:
"An attempt by any outside force to gain control of the Persian Gulf region
will be regarded as an assault on the vital interests of America, and such an
assault will be repelled by any means necessary, including military force."[10]
The doctrine declared the Persian Gulf to be internal US waters. Presidents
George H.W. Bush (1989–93) and George W. Bush (2001–09) used the Carter

doctrine to justify intervening in Kuwait in 1991 and illegally invading Iraq in 2003. Both were largely wars over oil. us-based corporations also enter other countries and become part of their internal ruling circles. As Harold Innis wrote, "American imperialism ... has been made plausible and attractive in part by its insistence that it is not imperialistic."[11]

Second Road to Energy Satellite

us petro-corporations have consistently pressed Washington to get first access to Canadian energy. They also influenced Canada from the inside, blending in as if they were Canadian and joining Canada's corporate leadership in ways managers of Chinese state oil corporations could not. "Foreign capital is able to determine possible governments by incarnating itself as an indigenous ruling class," George Grant wrote. By 1961, Diefenbaker's National Oil Policy was copying us energy policies. To boost its high-cost domestic industry, President Eisenhower placed quotas on importing oil and raised its domestic price above the international price for most of the us market. Washington treated oil from Western Canada, which was also a high-cost region, as if it were American, exempting Canadian oil from import quotas in exchange for Canada agreeing not to build an oil pipeline (Interprovincial Pipeline) to Montreal (the later Enbridge Line 9). The protection of us oil from cheaper imports briefly departed from using foreign oil and saving us oil for later.[12]

Trudeau's government followed Washington's precedent of holding a domestic oil price that varied from the international price. The difference was that Canadian oil prices were below rather than above the international price. The outcry from Alberta was deafening, though there had been silence from the same quarter in the 1960s, when Ontarians paid a third more than the international price to support Alberta's oil industry. But Canada's laissez-faire, continentalist oil policies that so favoured big foreign oil in the 1960s could not deal with the 1970s international oil crises. Canada quickly changed course and joined the resource nationalism trend that was spreading around the world.

Brief Energy Independence

Coming off the high of the 1967 centennial celebrations, Canadians grew confident they could find a different path from that of the us and gained much energy independence in the 1970s and early 1980s. Many saw the us in an increasingly negative light, as a result of the Vietnam War, race riots, guns, and the assassinations of John F. and Robert Kennedy and Martin Luther King.

In contrast, many defined Canada as a peacekeeping, caring-sharing society typified by medicare, multiculturalism and an inclusive ethos.[13] Many Canadians embraced economic nationalism, including Canadian ownership of domestic resources.

Pushed by the left and citizens' movements, Trudeau's government adopted Canada-first energy policies that culminated in the 1980 National Energy Program. The NEP rode a wave of anger at foreign oil taking huge profits while motorists paid two or three times as much for gasoline. In 1972, before the first oil supply crisis, Big Oil in Canada earned $18 in net income for every $100 earned by all other non-financial industries. By 1980, Big Oil's take had jumped to $42 dollars per $100.[14] The NEP promised Canadians energy security and a share of oil's super profits.

Key goals included ending foreign oil domination, promoting private Canadian ownership and rapidly expanding Petro-Canada, the government oil company. Joining Britain and France's oil nationalizations, Petro-Canada bought several foreign-owned oil transnationals.[15] Established in 1976, Petro-Canada was Canada's flagship, acting "for the people" through conservation, bringing Eastern Canadians energy security by reducing oil imports, moving Canadians off oil while finding new oil supplies, and providing a public "window" on the secretive industry.

While many countries reached 100 percent domestic ownership through nationalization, Canada's NEP set a modest goal of 50 percent Canadian ownership in ten years. Tax incentives and exploration grants would boost private Canadian ownership, and Petro-Canada would buy several large, foreign oil firms. It's rare that 84 percent of Canadians support anything, but 84 percent backed Canadianization in 1981. Two-thirds wanted to go faster and higher and wanted Petro-Canada to buy a big foreign company: Imperial, Gulf, Shell or Texaco. Even a majority of Albertans supported Canadianization. Harvie André, a Tory energy critic from Calgary and a fierce NEP foe, acknowledged that "the NEP is a successful policy. The people like it." Petro-Canada's advertising slogans reflected many Canadians' enthusiasm: "Buy Petro-Canada and pump your money back into Canada"; "Petro-Canada: It's Ours"; and "Canada First." Ottawa added a four-cents-a-litre gasoline tax to fund Petro-Canada's oil buyouts. Canadianization was quickly successful, rising from 7 percent of oil production ownership to 48 percent by 1985. The NEP almost reached its goal of 50 percent before Mulroney's Conservatives ended it.[16]

Canada became much more oil self-sufficient. In 1976, the Interprovincial Pipeline (today's Enbridge Line 9), a Sarnia-to-Montreal pipeline, brought

Canada's Two Roads to Becoming a US Energy Satellite

US CORPORATE OWNERSHIP AND CONTROL

CANADIAN ENERGY POLICIES THAT PRIORITIZE EXPORTS TO THE US

1960
90 percent of Canadian oil and natural gas is under foreign ownership and control.

1899
US-owned Standard Oil buys Canadian-owned Imperial Oil.

1940s
US conserves domestic supplies and pursues cheap foreign oil.

1961
Canada copies US energy policies and announces its own National Oil.

early 1970s
Canada promotes economic sovereignty and energy independence into the 1980s.

1950s
US turns to Canada for stable oil supply.

1976
National oil company Petro-Canada begins operations. Oil exports to the US are redirected to Quebec.

1959
US exempts Canada from oil quotas in exchange for not building an oil pipeline to Montreal.

1980
US dependence on foreign oil leads to the Carter Doctrine: US National interests in the Persian Gulf will be defended militarily.

1980
The National Energy Program (NEP) promises energy security and a share of oil's super profits.

1981
84 percent of Canadians support nationalization of the energy sector.

1981
US vigorously attacks Canada's National Energy Policy, threatening retaliation and economic destabilization.

1985
Pressure from Big Oil, petro-elites, the US and Western Canadian opposition ends Canada's NEP.

1991
US uses the Carter Doctrine to justify Kuwait intervention.

1987
Canada/US FTA gives the US proportional access to Canadian

2003
US uses the Carter Doctrine to invade Iraq.

1994
Canada, US and Mexico sign NAFTA.

Western Canadian oil to Quebec for the first time. Canada freed up oil for Montreal by cutting oil exports to the US. Ottawa promised to end all oil exports by 1982. The combined effects of the Interprovincial Pipeline, conservation measures and an off-oil program cut oil imports to less than 30 percent of their 1973 level.[17] But Ottawa faced a fairness challenge. If Canada imported oil in the east, how could Canada have a uniform oil price lower than OPEC's cartel price? Simple: Ottawa used revenue from a tax on oil exports from Alberta to the US to subsidize lower oil and gasoline prices for Eastern Canadians, who still bought foreign oil. Canadians paid the same lower price at the pumps wherever they lived. Easterners liked it.

Meanwhile René Lévesque's Parti Quebecois came to power in Quebec in 1976, promising a referendum on sovereignty-association. The sovereignty drive emboldened Trudeau on oil. Trudeau became prime minister in 1968 mainly to keep Quebec in Canada. He knew that showing Quebecers the economic benefits of staying in Canada would work better than talk of emotional bonds with other Canadians. Winning them to federalism was a major reason his government so daringly took on Washington, Alberta and the petro-giants. The NEP and Canadianization appealed more to Quebecers than to other Canadians.[18] Quebec had no known oil or natural gas then. Seeing images of Americans lining up at the pumps during the 1973–74 oil boycott, Quebecers felt secure getting Canadian oil and at a lower price. It was also appealing for Quebec to share in federal oil revenues. You won't get these benefits if you separate, Pierre Trudeau told them.

The NEP aimed to end oil imports, raise Canadian ownership and control of the oil industry, and set a made-in-Canada oil price.[19] Unprecedented oil revenues would be shared by all Canadians. The NEP's first two official goals were admirable and were supported by most Canadians. The revenue-sharing goal, though, roused the ire of Western Canadians, particularly Albertans, who saw it as encroaching on the provinces' resource powers and stealing their boom. Provincial control over resources had been contested since Saskatchewan and Alberta became provinces in 1905. Both provinces fought for twenty-five years to gain control over their resources, a right other provinces already had. Westerners have been suspicious of Ottawa's intentions to grab their resources ever since. The NEP's unofficial, swaggeringly partisan subtext particularly rankled. In *The Politics of Energy*, Canadian political scientists Bruce Doern and Glen Toner summarized the goals of the NEP as:

- restructuring political power between Ottawa, the oil industry and producing provinces (especially Alberta)
- reasserting federal powers over the economy and Ottawa's visibility in the hearts of Canadians
- getting greater federal revenues
- presenting the Liberals as the party that speaks for all Canadians[20]

While popular in Eastern Canada, the NEP was greatly resented by big foreign oil, Alberta's petro-elites and official Washington. They counter-attacked. After a burst of support for Canadianization by ordinary Albertans, a popular movement rose explosively in reaction, epitomized by the very popular bumper sticker, "Let the Eastern bastards freeze in the dark." Most Albertans saw the NEP as an unfair federal attack on Alberta's control over its energy and wealth. Many rallied across the province; 2,700 filled Edmonton's Jubilee auditorium, rising to their feet and chanting "Free the West" after rabble-rouser Doug Christie recited a litany of ill-treatments by central Canada. The *Edmonton Journal* was flooded with calls from those wanting to join Western separatist parties.[21]

"The NEP only reinforced the popular belief," wrote Alberta oil expert Larry Pratt, "that central Canadians continue to regard the west as little more than a resource-rich hinterland, and Alberta in particular as a storehouse of energy resources which are too important to be left under provincial jurisdiction."[22] The Trudeau government's tough unilateral opening position, meant to bring Alberta to the table, outraged many Westerners and failed at first to get an oil and revenue agreement with Alberta. So had the previous Joe Clark federal government (1979–80), even though Clark hailed from Alberta and the same Conservative party as Peter Lougheed. Lougheed's government didn't support Western separatism, but asserted provincial control over resources by cutting Alberta's crude oil production by 180,000 barrels a day, forcing oil imports to rise and undermining the NEP's oil self-sufficiency goal.[23] That brought Trudeau to the table and an Alberta–federal agreement was reached.

Despite the NEP's egregious faults on provincial resource control, it deserves credit as the only time Ottawa ever stood up to Big Oil and its allies in Washington. The provinces also took bold action at the time. Canadians were feeling feisty. Quebec nationalized US defence giant General Dynamics and its asbestos subsidiary in 1981. Saskatchewan held the world's best potash plays, but they were overwhelmingly foreign-owned, leaving Saskatchewanians mainly as diggers and shippers of potash. Allan Blakeney's

NDP government won the 1978 provincial election promising a government takeover of 40 percent of the industry and shifting higher-order activities, including head office jobs, to Saskatchewan.

The NEP might have succeeded, but its timing was off. The Global North's strong conservation measures to counter OPEC's power worked a little too well. World oil demand fell considerably for the first time in history. The world oil price crashed in 1982 and again in 1986. All of the world's oil-producing regions suffered devastating busts. Some oil-exporting countries in the Global South that borrowed heavily during boom times tottered. Mexico defaulted on its international debt. The US Federal Reserve Bank greatly accentuated the debtor countries' plight by pushing the US bank prime rate to an astonishing 21.5 percent. Other countries followed suit, debtors couldn't repay loans and the world economy quickly tanked. Canadian capitalists who took up the NEP's Canadianization incentives to buy foreign oil assets faced burdensome interest-carrying charges.

Bad timing also saw the surprise election of Ronald Reagan a week after Ottawa launched the NEP. Big Oil backed Reagan's election. He was very hostile to the NEP's ideological challenge that contradicted his neoliberal philosophy of small government and deregulation. The Canadian challenge was right in America's backyard.[24]

Even while Carter was still president, US officials quickly flew to Ottawa to state their shock at and disappointment with the NEP. The next summer, Reagan's government threatened Canada with retaliation and economic de-stabilization. One US official told a Canadian official, "If we pass the word to Wall Street that Canada is not a good place to invest, the [Canadian] dollar will fall, inflation rates will rise and the standard of living will fall."[25]

Surprisingly, Washington was more incensed by measures to increase private Canadian ownership than by the expansion of government-owned Petro-Canada. Executives from middle-sized US oil companies complained about Canada's "discriminatory" penalties on foreign corporations and incentives for Canadian-owned companies. Big Oil put up a loud fuss, but did well out of the NEP. Larry Pratt argues, "The NEP's structure of land tenure and incentives inadvertently encouraged the multinational oil companies to shift much of the risk and cost of frontier exploration to their Canadian partners and the Canadian state…The main beneficiaries of the National Energy Program…were the multinational oil companies."[26] None left Canada, and none would sell its shares to Petro-Canada, to the latter's regret.

Contrary to folklore, the NEP did not push Alberta into recession in the early 1980s, according to Frank Atkins, conservative economist and MA thesis advisor of Stephen Harper at the University of Calgary. Alberta did not use its surplus to counteract the slowdown.[27] Recession hit Alberta only in 1986—after Trudeau retired and the NEP was gone. We bought a house in Edmonton in early 1982, so I could start teaching at the University of Alberta and moved my young family from Ontario. It was the end of the oil boom, when house prices were highest. Four years later, our house had lost a quarter of its value, while house prices were skyrocketing in other provinces. My unease did not compare to the pain many Albertans felt in losing their jobs, businesses and houses. Unemployment hit 15 percent in Edmonton in 1986. People were so desperate that two thousand applied for a few low-wage jobs on Gainers Meats' kill-floor.

Alberta's media and Tory government hysterically blamed the NEP for Alberta's economic woes. While sensational "oil glut" headlines were seen in the *New York Times*, Alberta's headlines screamed "NEP." In Louisiana, my brother-in-law Fran was laid off as a diver for the oil industry. So were thousands of others in US oil-dependent states. The oil industry and its workers were down everywhere in the world, and as badly as in Alberta. Did Canada's NEP cause that, too? It was surreal to hear Calgary's oil executives and federal and Alberta Conservatives insist that Alberta's economic devastation was caused by Ottawa rather than by the market. Hadn't they claimed a few years earlier that the high international oil price resulted from supply and demand? Big Oil and Alberta's government were entitled to those high oil prices and revenues, they insisted. OPEC's cartel role in restricting oil supplies to drive up oil prices was ignored. But when international demand fell in 1986 and oil prices dropped by 75 percent, Trudeau was blamed.[28]

Brian Mulroney's Conservatives won a stunning victory in 1984, the election after Trudeau retired, partly on an anti-NEP wave. To put a stake through the NEP, Mulroney's Conservatives, backed by Calgary's petro-elites, sought an energy deal with the US that they called a "free trade" agreement. Proportionality was put in it to stop future NEPs. Fear of Alberta's reaction is why, when the old NEP was killed, it was replaced by *no energy policy*. The NDP government in Alberta, though, presents an opening for change. Premier Notley has said that Alberta wants to join a national plan to combat climate change. Will enough Albertans reject the view that what's good for Big Oil is good for them, so that Alberta can start down the road to a low-carbon future?

Canada's New NEP: No Energy Plan

Vancouver e-journalist Dan Crawford phoned Natural Resources Canada (NRCan) in 2009 to get its plan on oil shortages. An NRCan media relations person assured Crawford there is no plan because "it's impossible." Is it impossible because it's too complex to implement, Crawford asked. "No, it's impossible because a disruption could never happen," was the reply.[29] That's not NRCan's official response, but it reflects its mindset. Shane Mulligan, a University of Waterloo post-doctoral fellow, asked NRCan in 2009 about the Energy Supplies Allocation Board, which a 1985 Act of Parliament set up to assist in determining rationing and allocating energy resources during energy emergencies. Mulligan found what energy security researchers long suspected: the Energy Supplies Allocation Board does not exist. He was told it is a "virtual board" that will be struck after an oil shortage causes a national emergency. Four years later, in 2013, Director of Oil Sands and Energy Security at NRCan Doug Heath confirmed that the Energy Supplies Allocation Board did not meet.[30]

NRCan's reckless complacency is based on blind market faith, combined with a naive assumption that Canada has unlimited resources. Canada is "the most energy-secure nation in the world," largely because of the oil sands, asserted Ed Stelmach when he was Alberta's premier. Another former Conservative Alberta premier, Ralph Klein, was also misinformed. He pooh-poohed a Parkland Institute report warning that increasing exports of "Canada's oil and natural gas to the US puts our own energy security... in jeopardy." "There's a three-hundred-year supply of oil predicted in the Tar Sands," Klein quipped. Gary Lunn, then Canada's Natural Resources Minister, also rejected a Parkland report I wrote calling on Canada to set up strategic petroleum reserves. We have "close to two hundred years of current domestic demand" in crude oil reserves, Lunn stated. When other countries use their oil reserves in an international shortage, Canada could import the surplus freed up. This logic is like claiming not to need insurance on your house because neighbours have insurance on theirs. [31]

It is naive to assume that Canadian and international opinion will continue to allow the Sands to emit so many greenhouse gases. Or that Sands oil will save Eastern Canadians from oil shocks. Imports supply 8 percent of Ontarians oil, 80 percent of Atlantic Canadians oil and over 90 percent of Quebecers oil. Most oil pipelines from Western Canada point due south. As well, between 40 and 60 percent of Newfoundland oil is exported.[32] In 2011, 73 percent of Canada's total oil output was exported, almost all of it to the US.[33]

More of Canada's oil exports will head overseas if pipelines to the coast, any coast, are built.

Canada's plan for oil shortages is demand restraint, surge production and hope that the inventories private oil companies normally keep on hand will tide us over. Ottawa has to declare a national emergency before it can curb demand and determine who gets gasoline. First in line is "health, welfare and the security" of Canadians—hospitals, fire, police, national defence or public transit. Second is "economic stability"—public utilities, postal services, residential heating, taxis, road maintenance, and most industries and commerce. At the rear are "discretionary activities"—driving to work or the store.[34] These are the right priorities if we have to limit demand, but these extreme measures would not be needed if Ottawa ensured that Canadians, not the US, have priority access to domestic conventional oil.

Surge production can't help much. Following Hurricanes Katrina and Rita in Louisiana in 2005, Alberta raised oil output by thirty thousand barrels a day to send oil to the US. It was a nice gesture, but was a measly 1 percent of Canada's daily production. Prolonged surges damage oil fields. Furthermore, 87 percent of the International Energy Agency's response to Hurricane Katrina came from SPR oil releases. Only 2 percent came from curbing demand. Canada pins its hopes on the 2 percent solution.[35] Federal Resources Minister Lunn explained in 2008 that while Canada does not mandate private industry to hold inventories for security purposes the way many European countries do, "in an emergency Canada has the power to mandate the building and use of crude oil and petroleum products inventories."[36] Oil corporations keep stocks for commercial operational purposes. Ottawa hopes they will somehow provide sufficient oil in an emergency. They won't. As we'll see in chapter 3, refiners in Eastern Canada hold nowhere near the IEA standard of ninety days' worth of oil imports.

While the reversal of Enbridge's Line 9 to a west-to-east direction is scheduled to begin sometime in 2015, it will carry substantial amounts of US shale oil because it gets to southern Ontario from Western Canada via North Dakota, where most of the Bakken shale oil field is located. We've seen that Washington would likely stop US oil from entering Canada in an international oil crisis, precisely the time when it would be needed. TransCanada's proposed Energy East oil pipeline has the advantage of an all-Canadian route that would not pick up US oil on its way from west to east, but it is not yet in place and is heavily contested. Will it ever be converted and built? Even if it is, most of the oil it would carry will be for export.[37]

Before Energy East is in place, Canada will not be able to easily ship domestic crude and refined oil to Eastern Canada. Some can be sent by rail, tanker trucks and Great Lakes ships, but they cannot carry enough. Another way is by ocean tanker from Vancouver going through the Panama Canal to Atlantic Canada and Quebec, but that is too slow and costly. An IEA report on Canadian oil security notes that the central provinces' geographic isolation makes them vulnerable to pipeline disruptions.[38]

Canada's refineries are not geared well to accept domestic grades of oil. While Newfoundland produces about enough crude to meet current demand in all of Atlantic Canada, about half is exported.[39] The Irving refinery in New Brunswick can handle Newfoundland's sweet light crude, but Newfoundland's sole refinery at Come By Chance processes only a small amount.

Canada is the only IEA member with growing oil output. If world oil supplies fall by 7 percent or more, Canada is expected to export more oil to other IEA members—meaning the US—rather than reduce exports in order to divert some of them to Eastern Canadians.

Oil Disruptions

Few Canadians know life without oil. Only those older than sixty lived when most Canadian families didn't own one car, let alone two. There were no oil shortages in Canada in 1973–74 like there were in the US. But hardships during events like ice storms give us an inkling of what prolonged oil scarcities could look like in winter. Torontonians, Quebecers and New Brunswickers suffered a blackout that lasted up to ten days during an ice storm and cold snap at Christmas in 2013. And the Montreal ice storm of January 1998 was a particularly Canadian nightmare.

No electrical power for up to several weeks meant no heat for more than four million people during the year's coldest month. Six died from hypothermia, and another six from carbon monoxide poisoning through faulty use of generators, but the death toll could have been much higher. Canada's economic output dropped 0.7 percent in the month. The storm revealed Quebecers' reliance on electricity, and impacted energy and clean water distribution, heating, hospitals and other health services, food output, telecommunications, transportation and banking.[40] Quebec set up a coordination centre to prioritize, organize and distribute power generators. Hospitals and other medical centres topped the list. Elective surgeries stopped. Some staff brought family to live in hospital with them so they could be available at all times. Roads were hazardous, and telephone and radio service was poor. Water pumps stopped working.

Quebec's main focus was to increase energy supply, a sensible thing to do when it runs short. This is not part of Canada's plans for oil and natural gas shortages, though. The focus is only on who gets reduced supply.

Other Energy Shortages

Cape Breton Islanders ran short of furnace oil on December 18, 2007. It was a dress rehearsal for more frequent and longer interruptions in Atlantic Canada, Quebec and parts of Ontario. "An early winter and a late-arriving fuel tanker have revealed to many Cape Bretoners a precarious home heating oil supply," the CBC reported.[41] "People used up their furnace oil quickly because winter came early this year," an Imperial Oil spokesperson stated.

The closest some Canadians have come to gasoline and diesel scarcities recently were shortages due not to crude oil shortages but to unplanned refinery shutdowns in Ontario in 2007 and 2011, Alberta in 2008 and 2013, and Nova Scotia in 2011. In contrast to the US, in Canada companies need not publicly admit supply interruptions. Refiners regularly schedule maintenance closures by storing fuel beforehand, but unexpected shutdowns bite because extra supply is not on hand. A gasoline shortage hit the Prairie provinces in 2013 after Suncor's Edmonton refinery needed unexpected repairs. Albertans had run short for three weeks in 2008 because of a malfunction at the same refinery. The irony of gasoline shortages in oil-rich Alberta was not lost on thousands of frustrated drivers. Building a new refinery in Alberta is more expensive than shipping Alberta oil to existing refineries in Texas. But these minor inconveniences did not budge Canadians' complacency about energy availability. Britain was better off for having gone through the petrol famine of 2000. Canadians may have to learn the hard way.[42]

Running Out of Easy Oil

"Peak oil" burst onto the scene in 2000 when Colin Campbell founded the Association for the Study of Peak Oil. Campbell is a venerated petroleum geologist who worked for Oxford University and British Petroleum and advised national governments. He and his associates warned that the era of cheap and plentiful oil is over. At first, they were widely viewed as kooky. Then, from 2005 to 2010, credible voices spoke up for peak oil and it became respectable.[43] But the US shale oil revolution, starting in 2011, pushed peak oil to the fringes again.

Leonardo Maugeri's 2012 report, *Oil: The Next Revolution*, shifted public expectations about when world oil will peak. Published by Harvard's Kennedy School, the report boldly asserts, "Contrary to what most people believe, oil supply capacity is growing worldwide at such an unprecedented level that it might outpace consumption. This could lead to a glut of overproduction and a steep dip in oil prices." Maugeri foresees additional world production coming from Iraq, US shale, Canada's Sands and Brazil's deep sea. The shale oil boom is not temporary and global conventional oil is growing, Maugeri asserts. After 2015, there could be oil overcapacity. Maugeri's report triggered a landslide of anti–peak oil opinion. His message was what the oil industry, stock markets, governments and mainstream media wanted to hear. The world oil glut and price crash in the fall of 2014 appeared to confirm his view. The new consensus is that the "peak-oil concept is increasingly out of date less than a decade after its proponents said global output would surely hit the halfway mark."[44]

The IEA's 2012 *World Energy Outlook* (WEO) echoed Maugeri's oil optimism. The respected WEO report stated that "a new global energy landscape...is being redrawn by the resurgence in oil and gas production in the United States." The 2013 WEO went further, predicting that the US would become the world's largest oil producer by 2015, with much of the gains from natural gas liquids. These predictions were subsequently borne out, but the media exaggerated them with headlines like "US Is on Fast-Track to Energy Independence" and America's "game-changing" prospect of a "hundred years of natural gas."[45]

The WEO predicts that newly found crude oil will counterbalance declines. "In 2000–11, fourteen billion barrels on average were found a year; the amount that needs to be discovered (and developed) to meet the Current Policies Scenario projections for 2012–35 is only seven billion barrels per year... Our projections allow for the fact that discoveries are projected to continue to decline over the longer term."[46] The IEA expected oil production to rise from 87 million barrels per day in 2012 to 98 million in 2035, with gains mainly coming from North American unconventional oils (including Sands oil), natural gas liquids and OPEC.[47] After 2019, when US shale oil begins to decline, the IEA and Washington are counting on Iraq to make "the largest contribution by far to global oil supply growth" for the next twenty years. How mad is that?

For two decades, Washington has hoped its invasions and bombings of Iraq will produce more oil in US-friendly market conditions. It's an idle hope. Iraq's oil production did not surpass its 1989 peak, before the first Gulf War, until 2011. Although oil output was up considerably by 2015, the Islamic State of Iraq and the Levant (ISIL) insurgency threatens these gains. The IEA sees a downside to its predicted growth in oil supplies. Carbon emissions from future oil projections correspond to a global temperature rise of 3.6 degrees Celsius, almost double the rise scientists deem acceptable.[48]

With all the evidence, is there room for skepticism that peak oil is now a distant concern? Yes, plenty. Both Maugeri and the IEA have a vested interest in making overly optimistic assumptions about the deliverability of potential oil supplies. Meanwhile, the less exciting news of major declines in large old oil fields and rapid depletion rates in shale oil plays are downplayed. The law of diminishing returns, political instability in some major oil-producing countries and government measures to combat climate change are underrated.

Maugeri has an oil-industry bias. A top manager at ENI, Italy's oil and natural gas giant, Maugeri drew on his proprietary database developed at ENI for his study and thanked BP for funding it. Jean Laherrère, a petroleum engineer for thirty-seven years with Total, France's oil giant, castigated Maugeri's report. He notes that Maugeri's projections for Alberta's Sands exceed those of the usually overly optimistic Canadian Association of Petroleum Producers. Likewise, Maugeri's US oil output projections surpass those of the US government. Finally, Maugeri relies on "proved and probable" reserves that in the past were higher than actual reserves. These flaws discredit Maugeri's projections, Laherrère contends. Maugeri cut by half depletion rates established in other studies for existing oil fields. By boosting production and lowering depletion forecasts, Maugeri arrives at rising world oil output.[49]

Are his assumptions credible? Oil corporations use extraction techniques like horizontal drilling to slow a reservoir's depletion rates, only to see huge plunges later, an *Oil Drum* article explains: "When the wells run horizontally at the top of the reservoir, they are no longer reduced in productive length each year as vertical wells are, because the driving water flood slowly fills the reservoir below the oil as it is displaced. This does not mean that…it will ultimately produce more oil."[50]

What about the IEA's forecasts? Most advanced countries are complacent about coming world oil shortages because they rely on IEA forecasts. The *Guardian* newspaper has cast doubt on the IEA's objectivity. A senior IEA official who requested anonymity said the world is much closer to running out of oil than the IEA admits. It has deliberately underplayed a looming shortage for fear of triggering panic buying, he charged. Also, the US encourages the IEA to underplay the decline rate from existing fields and overplays chances of finding new reserves.[51] A former senior IEA source said a key imperative at the agency is "not to anger the Americans." There is not as much world oil as had been stated. "We have [already] entered the 'peak oil' zone…. The situation is really bad," he added. The IEA rejected the whistle-blower's allegations.[52] To regain credibility, the IEA studied decline rates in 580 of the world's biggest fields that had passed peak production. Decline rates were almost double earlier estimates.[53]

For years, the IEA repeatedly pooh-poohed peak oil. Then its 2010 *World Energy Outlook* nonchalantly stated that "crude oil output reaches an undulating plateau of 68–69 [million barrels per day], but never regains its all-time peak of 70 mbd reached in 2006," taking peak oil from unworthy of discussion to already in the past. Former White House energy advisor Matt Simmons contended that global conventional crude oil actually peaked in May 2005, but the peak was masked by temporary boosts in petroleum liquids, largely from natural gas, refinery processing gains and inventory draw-downs.[54]

In 2013, Michael Höök, Robert Hirsch and Kjell Aleklett re-estimated declines in the world's major oil fields:

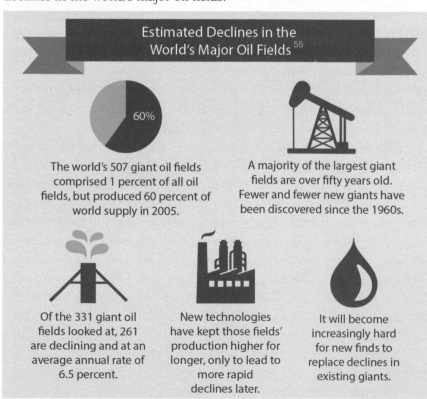

Estimated Declines in the World's Major Oil Fields [55]

60%

The world's 507 giant oil fields comprised 1 percent of all oil fields, but produced 60 percent of world supply in 2005.

A majority of the largest giant fields are over fifty years old. Fewer and fewer new giants have been discovered since the 1960s.

Of the 331 giant oil fields looked at, 261 are declining and at an average annual rate of 6.5 percent.

New technologies have kept those fields' production higher for longer, only to lead to more rapid declines later.

It will become increasingly hard for new finds to replace declines in existing giants.

Peak-oil expert Gail Tverberg deems the IEA's oil output estimates excessive. As easy-to-produce oil is depleted, oil corporations move to more arduous fields, whose extraction costs are higher. New American shale oil will stay on-stream and the US can become the world's leading producer, Tverberg contends, only if oil prices soar more than the IEA forecasts.[56]

Some auto and oil executives get it. "Our view is that oil production will peak in the near future" stated Katsuaki Watanabe, Toyota's vice chairman. "We need to develop power train(s) for alternative energy sources." Jim Lentz, CEO of Toyota North America, foresees oil peaking by 2020. Christophe de Margerie, who until 2014 headed the giant oil corporation Total Petroleum, insisted that oil will soon top out at 95 million barrels per day and no amount of investment or ingenuity can push it higher. "There will be a lack of sufficient energy available." Echoing Maugeri, de Margerie sees peak oil coming from political resistance: "The problem is not with resources, it is how to extract resources in an acceptable manner." Politics stand between large geological supplies and peak deliverability. Bans on fracking, carbon pricing, cap-and-trade plans and other actions to reduce emissions and protect habitats will reduce deliverable supplies. So can nationalizations, popular resistance, civil wars, decisions to leave oil in the soil, and high production costs in the Sands, the deep ocean and the Arctic combine to hinder new oil output.[57]

The oil price crash of 2014, triggered by Saudi Arabia's refusal to cut OPEC's oil output, seemed to support Maugeri's predictions. By 2015, talk of peak demand had replaced that of peak oil. Demand is already falling in developed countries because of a switch to natural gas and major strides in vehicle efficiency. More efficient autos and factories are greatly slowing oil demand growth in China. These trends worry Ali al-Naimi, Saudi Arabia's petroleum minister, who opposes action on climate change. Saudi Arabia will stand "firmly and resolutely" against attempts to marginalize oil consumption. "There are those who are trying to reach international agreements to limit the use of fossil fuel," al-Naimi declared. That will damage oil producers' interests. Al-Naimi wants to replace the American fixation on "security of supply" with Saudi's need for "security of demand."[58] Saudi Arabia's stance confirms Maugeri's claim that oil is not physically in short supply, but that oil's fortunes are political.

Oil Peak Year and Quantity Compared to Declining Production in 2010[59]

Thousands of barrels of oil produced per day

USA	1970 = 11,297 / 2010 = 7,513 (−33.5%)
Venezuela	1970 = 3,754 / 2010 = 2,471 (−34.2%)
Other Middle East	1970 = 79 / 2010 = 38 (−52.2%)
Libya	1970 = 3,357 / 2010 = 1,659 (−50.6%)
Kuwait	1972 = 3,339 / 2010 = 2,508 (−24.9%)
Iran	1974 = 6,060 / 2010 = 4,245 (−30.0%)
Romania	1976 = 313 / 2010 = 89 (−71.5%)
Indonesia	1977 = 1,685 / 2010 = 986 (−41.5%)
Trinidad & Tobago	1978 = 230 / 2010 = 146 (−36.6%)
Iraq	1979? = 3,489 / 2010 = 2,460 (−29.5%)
Brunei	1979 = 261 / 2010 = 172 (−34.0%)
Peru	1980 = 196 / 2010 = 157 (−19.8%)
Tunisia	1980 = 118 / 2010 = 80 (−32.7%)
Other Europe & Eurasia	1983 = 12,938 / 2010 = 374 (−97.1%)
Other Africa	1985 = 241 / 2010 = 143 (−41.0%)
Russian Federation	1987 = 11,484 / 2010 = 10,270 (−10.6%)
Egypt	1993 = 941 / 2010 = 736 (−21.7%)
Syria	1995 = 596 / 2010 = 385 (−35.4%)
Gabon	1996 = 365 / 2010 = 245 (−32.8%)

Legend:
- Peak Production
- 2010 Production
- % Off Peak in 2010

X-axis: 0 1 2 3 4 5 6 7 8 9 10 11 12 13

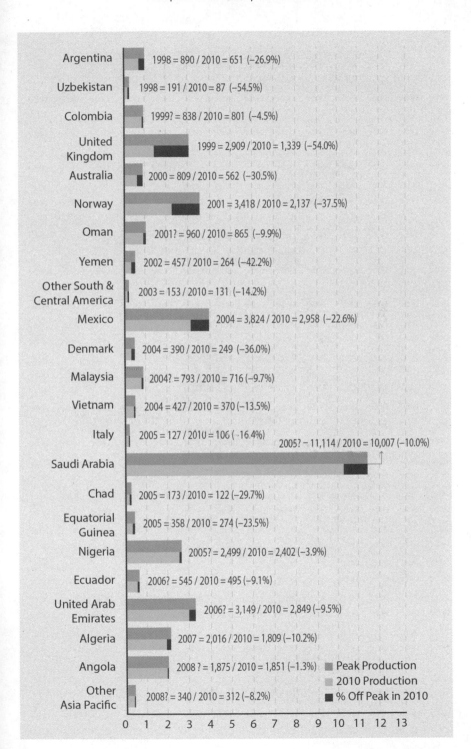

Argentina 1998 = 890 / 2010 = 651 (−26.9%)

Uzbekistan 1998 = 191 / 2010 = 87 (−54.5%)

Colombia 1999? = 838 / 2010 = 801 (−4.5%)

United Kingdom 1999 = 2,909 / 2010 = 1,339 (−54.0%)

Australia 2000 = 809 / 2010 = 562 (−30.5%)

Norway 2001 = 3,418 / 2010 = 2,137 (−37.5%)

Oman 2001? = 960 / 2010 = 865 (−9.9%)

Yemen 2002 = 457 / 2010 = 264 (−42.2%)

Other South & Central America 2003 = 153 / 2010 = 131 (−14.2%)

Mexico 2004 = 3,824 / 2010 = 2,958 (−22.6%)

Denmark 2004 = 390 / 2010 = 249 (−36.0%)

Malaysia 2004? = 793 / 2010 = 716 (−9.7%)

Vietnam 2004 = 427 / 2010 = 370 (−13.5%)

Italy 2005 = 127 / 2010 = 106 (−16.4%)

Saudi Arabia 2005? − 11,114 / 2010 = 10,007 (−10.0%)

Chad 2005 = 173 / 2010 = 122 (−29.7%)

Equatorial Guinea 2005 = 358 / 2010 = 274 (−23.5%)

Nigeria 2005? = 2,499 / 2010 = 2,402 (−3.9%)

Ecuador 2006? = 545 / 2010 = 495 (−9.1%)

United Arab Emirates 2006? = 3,149 / 2010 = 2,849 (−9.5%)

Algeria 2007 = 2,016 / 2010 = 1,809 (−10.2%)

Angola 2008 ? = 1,875 / 2010 = 1,851 (−1.3%)

Other Asia Pacific 2008? = 340 / 2010 = 312 (−8.2%)

■ Peak Production
■ 2010 Production
■ % Off Peak in 2010

0 1 2 3 4 5 6 7 8 9 10 11 12 13

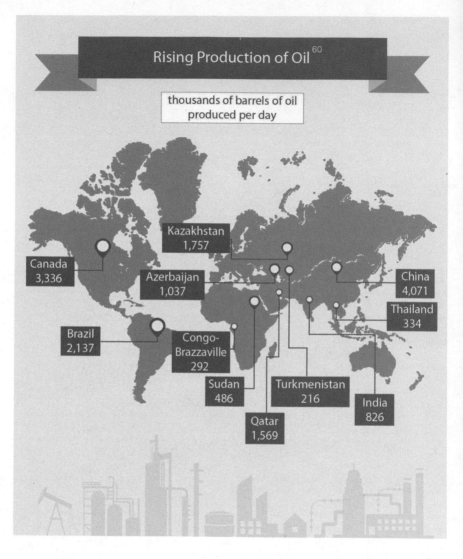

Rising Production of Oil[60]

thousands of barrels of oil produced per day

Kazakhstan 1,757

Canada 3,336

Azerbaijan 1,037

China 4,071

Thailand 334

Brazil 2,137

Congo-Brazzaville 292

Sudan 486

Turkmenistan 216

India 826

Qatar 1,569

However, it's not a question of peak demand or peak oil. Both trends are happening and they're both good news for the environment. Both are limited by politics. That means citizens movements and government actions make a difference. Although it avoids the toxic phrase *peak oil*, Shell predicts supply shortages for conventional oil and natural gas by 2020.[61] Because oil executives are free to speak openly after retiring, former Shell Chairman Lord Ron Oxburgh wrote that "there isn't any shortage of oil, but there is a real shortage of the cheap oil... It is pretty clear that there is not much chance of finding any significant quantity of new cheap oil." State oil companies control

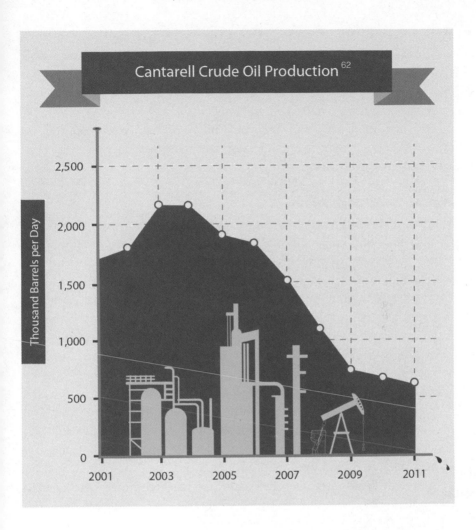

Cantarell Crude Oil Production[62]

80 percent of the world's oil and gas reserves and have legitimate policies different from the oil majors, Oxburgh explains. It's reasonable that they should leave it in the ground for their children."[63]

Mexico's Cantarell oilfield shows how quickly giants can fizzle. Once the world's third largest field, by 2012 it had plunged more than 80 percent from its 2003 peak. Its fall means Mexico will likely go from major crude oil exporter to net oil importer by 2020. A study based on BP's *Statistical Review of World Energy* found that three times as many oil-producing countries were in decline (thirty-seven) as countries where it was rising (eleven). To be a decliner, a country must be 10 percent below peak, and its peak must have occurred more than five years in the past. Canada is a riser. Saudi Arabia is a question mark. In 2010, it produced 10 percent less than its peak in 2005, but its output was up by 2015 and its government claims to have ample spare capacity.[64] If Saudi Arabia is moved into the growth camp, it would appreciably shift the oil output percentage to that camp. But official Saudi reserves and output forecasts are dubious.

Oil output is definitely growing only in the bottom eleven countries on the list. Will shale, tight oil and fracking push some declining countries back into the growing column?

History of Oil Discoveries

The two largest oil fields ever found were Burgan in Kuwait (1938) and Ghawar in Saudi Arabia (1948). Much of the world's current supply still comes from fields discovered before 1970, including these two. Super giants are generally the "easiest to find, the most economical to develop, and the longest lived." Finding new super giants is very unlikely. The exception is perhaps the deep ocean. Petrobras, Brazil's government-owned oil company, is the world leader in ultra-deep-sea wells. The Lula field, seven kilometres below sea level, holds 8.3 billion barrels of oil. Carcara, another deep-sea sub-salt field discovered in 2013, holds up to 33 billion barrels. Mainly because of the deep-sea fields, the IEA expects Brazil's oil output to reach 6 million barrels per day by 2035, and provide a third of the world's net oil production growth by then. It sounds like a lot, but it would provide only about 6 percent of world output. Deep-water sub-salt exploration has also started in the US part of the Gulf of Mexico.[65]

Although these ultra-deep fields are big, they are much smaller than Ghawar and Burgan once were. They're insufficient to reverse peak oil world-wide. At $100–$150 million a well, deep-ocean fields cost three or four times more than traditional offshore wells. Their addition to world oil output will push the international oil price very high. Like unconventional shale oil and natural gas wells, deep-water oil fields usually reach peak quickly and steeply decline. For example, Thunder Horse, the largest production platform in the Gulf of Mexico, has been declining by about 4 percent a month. Even if new deep-sea super giants are found, they will use much more energy to extract and process than conventional oil. When net gains from using carbon energy (e.g., natural gas) to produce new carbon energy (oil) falls to two or three to one, they're not worth the effort.[66]

Official oil reserves are another problem. OPEC changed its quota system in the 1980s on the allowed oil output of member countries, partly basing them on a member's "proven" national reserves. Governments then inflated official reserves. If they admit that their country's oil reserves have declined greatly, petro-governments can collapse. In 1988, Venezuela doubled its official reserves. The United Arab Emirates, Iran and Iraq immediately double or tripled theirs. Two years later, Saudi Arabia raised its reserves by half.[67] It's hard to believe that huge, simultaneous discoveries occurred in all these countries and were unrelated to OPEC's new quota rules. Many national oil companies and governments in OPEC do not use external audits of their reserves and they hide results, claiming they're state secrets.[68]

Curiously, oil reserves in these countries have not fallen much after over twenty-five years of pumping. A cable from the US embassy in Riyadh released by WikiLeaks in 2011 warned that Saudi Arabia's oil reserves may be overestimated by as much as 40 percent. In 2006, Kuwait's oil reserves were revealed to be half those of its official numbers. Official reserves are "political reserves." In 2013, OPEC claimed to have 81 percent of the world's oil reserves. Astoundingly, the US Energy Information Administration and the International Energy Agency IEA accepted and repeated them.[69]

Some oil corporations also hold the equivalent of political reserves. Shell Oil was known for overcautiousness—until it was caught in 2004 overestimating its "proven" oil reserves by 4 billion barrels. Its share values fell by $4 billion in one day.[70] Corporations also have an incentive to inflate their reserve numbers. The main points to note are: 1) deliverability and political decisions can determine oil output more than total theoretical supply; 2) we have exhausted the easiest, cheapest sources and the rest are subject to diminishing returns; and 3) despite the temporary rise in US oil and natural gas liquids, the world supply of environmentally acceptable and deliverable oil is being depleted.

Oil Shock: What Could Happen

Oil provides 93 percent of the energy used in transportation.[71] There is currently no major alternative to it in that sector. That's why oil prices rise so much when supplies dip slightly. Demand for oil for transport is so inelastic that in the short-run, drivers pay what they must to get to work or buy groceries. Only after gasoline prices stay high for long periods do motorists switch to more efficient vehicles or public transit, walk, cycle or move closer to work. After the 1970s oil crises, there were massive shifts "off oil" by power utilities and for home heating. Fewer opportunities remain today to go off oil. Fleet vehicles in North America are switching to natural gas because temporary natural gas surpluses have driven down prices so much. The main switchers are line haul trucking, return-to-base trucks, transit buses, garbage trucks, taxis, courier vehicles, marine vessels, rail locomotives and forklifts.[72]

A 2005 US report, *Oil Shockwave*, warned that taking just 4 percent of oil off the world market would lead to a 177 percent price rise. If oil prices were $100 dollars a barrel, such a cut could push it to over $250 a barrel or over $9 a gallon ($2.38 per litre) of gasoline, the report noted. A bipartisan successor panel repeated the oil shockwave scenario in 2011 and found similar results. Tight supplies mean that small disruptions have big effects on price and availability. Disruptions can occur because of natural disasters or wars. In 2005, for example, Hurricane Katrina, followed closely by Hurricane Rita, cut world oil output by less than 2 percent. Rapid releases from the US Strategic Petroleum

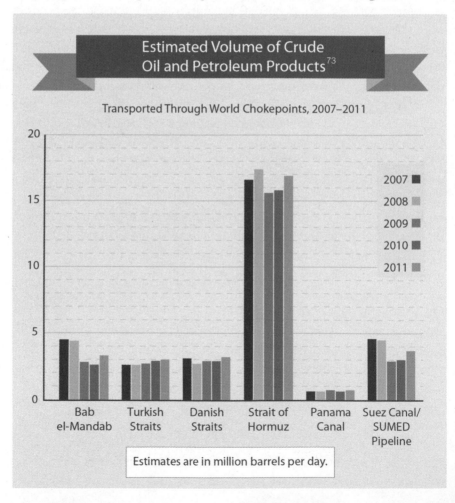

Estimated Volume of Crude Oil and Petroleum Products[73]

Transported Through World Chokepoints, 2007–2011

Estimates are in million barrels per day.

Reserve (SPR) averted a major price spike. In 2011, the IEA coordinated a release of oil from member countries' SPRs to counteract lost Libyan production when Western powers intervened in its civil war. Blake Clayton argues that tapping emergency reserves dampens oil prices temporarily but causes price rises soon after. Terrorist attacks or embargoes aimed at selected countries could lead to much greater supply cuts. *Oil Shockwave* looked at several scenarios: rising violence in Nigeria, explosions at a Saudi natural gas processing plant, and an attack on the oil port of Valdez, Alaska.[74]

Real events have paralleled these hypothetical disasters. Saudi oil facilities were targeted three times in 2006–07. Yemen foiled planned al-Qaeda attacks on its oil and gas infrastructure in 2013. Insurgents regularly blow up pipelines in Iraq, and ISIL controls or threatens some oil-producing regions. If Israel bombs Iran's nuclear facilities, Iran may try to close the Strait of Hormuz, the world's most strategically important chokepoint, at the mouth of the Persian Gulf. Inbound and outbound traffic lanes narrow to three kilometres. If blocked, 17 million barrels per day would be cut, about 35 percent of all seaborne traded oil and almost 20 percent of oil traded worldwide.[75]

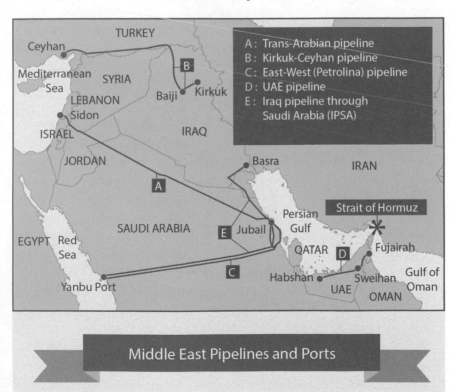

Middle East Pipelines and Ports

How would blockading Hormuz affect Canadians? In mid-2014, Canada got 11 percent of its oil imports from Saudi Arabia, Canada's number two source of oil imports, after the US.[76] That oil passes through Hormuz. Eastern Canadians couldn't easily replace that big of an import drop, because other countries would scramble for the same oil. It would be worse in winter, as many Atlantic Canadians, use oil furnaces. Asia would be hit hardest. Sixty-four percent of Japan's oil comes through Hormuz—and it's 23 percent for China, 71 percent for India and a stunning 92 percent for South Korea. But the whole world would be impacted, as the international oil price would skyrocket and food prices would soar.

Oil as a Political Weapon

Oil first became a potent weapon after the 1973 war between Israel, Egypt and Syria.[77] It was used effectively again during the 1978–79 Iranian revolution that overthrew the Shah, whom the US and Britain had imposed after they ousted the elected and popular President Mosaddegh because he nationalized their oil corporations. In the 1970s, bilateral, long-term contracts between exporting and consuming states, often negotiated by transnationals, dominated the world oil market. Oil was not very substitutable then, and countries were less able to deal with supply disruptions. If an oil-exporting country broke its long-term contract with a specific consuming country, the latter had difficulty finding alternative sources.[78] That is why oil embargoes were effective.

Things fell apart for oil boycotters in the 1980s, as oil use fell sharply and output rose in non-OPEC countries. Led by the International Energy Agency, industrial countries cut oil use more through fuel efficiency regulations and gasoline tax hikes than through market price rises. The world oil price dropped by 75 percent and remained low through the 1990s. The oil glut was also due to North Sea and Alaskan oil coming on-stream, greatly reducing IEA countries' reliance on OPEC oil. If OPEC countries had not nationalized Western oil transnationals, it's likely that North Sea and Alaskan oil would have developed much later.[79] The oil glut cut the ground out from under OPEC producers, who could no longer effectively boycott a specific country. These were perfect conditions for the neoliberal counter-revolution led by Conservative Margaret Thatcher in Britain and Republican Ronald Reagan in the US. Gutting environmental regulations during a world oil glut led to the creation of new exchanges, oil futures contracts and spot oil markets.

Oil is a generic product. Once it's on the global market it's interchangeable with oil of the same grade anywhere in the world. In the 1980s and 1990s, spot prices replaced long-term contracts. No country or petro-transnational could corner the oil market. If country A embargoed oil exports to country B, B could buy from countries C to Z. As oil analyst Joseph Stroupe put it, "Extremely liquid oil-futures contracts ('paper oil') that looked forward only a few months to a few years... and that could be freely and openly bought and sold on a daily basis on the new exchanges replaced the traditional, rigid, discrete long-term supply contracts negotiated directly between exporting and importing states. The global oil-market order was becoming tremendously liberalized, open and highly liquid under us leadership and control."

New oil exchanges led to a single "global pool of oil denominated in us dollars into which nearly all exporters sell their oil and out of which nearly all importers purchase oil." [80] The liberalized market undermined the power of oil exporters and targeted oil embargoes. The hand of the oil transnationals was strengthened. But things changed after 2000. The resurgence of state-owned oil companies turned private transnationals into rule-takers and supplicants to Russia and much of the Global South, from which most new oil supplies are expected to come. While the shale oil boom has shifted a chunk of the action back to the us, only one of the eleven countries where oil output is growing is in the Global North: Canada.

In the 1950s, the "Seven Sisters"—the world's giant, private oil corporations—controlled 85 percent of world oil reserves. Now national oil companies control 90 percent. [81] National oil companies are instruments of their government owners. The new big seven, in order of size, are Saudi Aramco, Gazprom (Russia), NIOC (Iran), ExxonMobil, PetroChina, Kuwait Petroleum and Shell. ExxonMobil and Shell's oil reserves of 25 and 8 billion barrels respectively are puny compared to Saudi Aramco's 307 billion and NIOC's 311 billion. [82] The petro-transnationals are declining because they are shut out of cheap oil and confined to-hard-to-get oil in the Sands, the deep sea and the Arctic.

Oil and natural gas are political weapons again. Iran has threatened to block Hormuz. Venezuela promises to embargo the us if it tries to overthrow its socialist government again. Russia has threatened to cut natural gas supplies to Belarus several times and did cut supplies to Ukraine. In the winter of 2007, Russia cut oil exports to Poland, Germany and Ukraine to prevent Belarus from accessing Russian supplies. With the exception of Iraq's INOC and Norway's Statoil, the big national oil companies are outside the us sphere of influence. They generally put the interests of their citizens, or more often of elites close to the government, above those of exports and other countries' interests. They engage in long-term contracts. National oil companies and boycott threats are combining to de-globalize oil.

Many adherents to the "magic of the marketplace" ideology seem not to have noticed that oil is reverting to 1970s conditions, but with twists. Higher oil prices and disputes with environmentalists and indigenous peoples over developing non-conventional oil and hazardous ways of moving it by train, pipe or tanker challenge Big Oil's once-easy dominance. China, India and Europe are getting more dependent on carbon fuel imports and more vulnerable to potential boycotts.

These are strong trends, but there are also countertrends. Surges in American natural gas and oil are making the US temporarily less dependent on imports and less vulnerable to embargoes. Nevertheless, rapidly changing conditions are undermining the Western-dominated corporate oil market. As oil supplies tighten, rising East and South Asian countries, anxious to secure long-term supplies, are bypassing the US-dominated spot markets in favour of long-term, state-to-state contracts with West and Central Asian oil producers. An Asian-centric system is displacing the US-centric one.

China is targeting Africa for oil and other resources because Western powers have not locked Africa up. Most African states are weak and lack national oil companies. China offers oil purchases, financial contracts, and investments in local roads and hospitals. China is taking more and more oil off the world market this way. Saudi Arabia is still China's largest supplier of oil imports, but Angola is now second. African opposition to Chinese domination appears to be growing, though. China is also turning to Russian oil. In the context of a fracturing global oil market, most countries are adopting national energy security strategies. Canadian authorities appear unaware of the new reality, leaving individual Canadians to fend for themselves.

Only Three Energy Security Plans, None Canadian

Rick Munroe, a farmer and former teacher living near Kingston, Ontario, researches peak oil and alerts usually incredulous government officials about coming oil shortages. In his polite, dogged way, he asks them for their plans for oil shortages. After long silences, most reply "we have none." Some point to emergency measures specifying who would get priority access to restricted supplies. None address how to replace imported oil. Munroe found only three plans for oil supply crises: Alan Smart's Tasman report for Australia, Kathy Leotta's Seattle-based study, and British government plans. [83]

The Tasman and Leotta studies show there will be chaos if people and local authorities are not well prepared long in advance. Without a publicly known plan, people will fend for themselves and hoard necessities, triggering shortages much sooner. Transport will quickly break down, causing scarcities of food, medicine, heating fuel and jobs. Both reports call for coordination among levels of government and agencies, new legislation and public education.[84] Governments in Australia and the us have ignored these sensible proposals. Britain, though, fashioned a good plan for disruptions to natural gas, electricity, oil and petrol after the 2000 fuel protest mayhem, delivering fuel to those most in need, protecting the economy and quickly restoring full energy supplies. There are detailed plans for many sectors and about eight hundred Designated Filling Stations to ensure fuel for first responders.[85] In a decentralized country like Canada, federal, provincial and local governments must learn to work together, not something they usually do well. And they must practise coordinating before a crisis hits. Canada has an Energy Supplies Emergency Act on the books, but ignores it. Even good legislation is not enough, though.

Chapter 3

Without a Parachute

Would Canada be impacted if you suddenly had an oil shortage? Sure. It would be shut down.

—Matt Simmons, chairman of Simmons International, one of the world's largest energy investment banks, 2008

It should be national policy to ensure that there is a minimum ninety-day supply [of oil] held in storage at all times.

—Robert Stanfield, Progressive Conservative, Leader of the Official Opposition, 1974

Strategic petroleum reserves (SPRS) are emergency pools of oil to be released during oil supply disruptions to replace temporary cuts to oil imports and protect residents' economic and physical well-being. The standard supply is ninety days of net imports. SPRS can be stored in salt caverns or above ground tanks. They can be held in government facilities, by special stocking agencies or by commercial oil companies that are required to stock a set amount for national emergencies. SPRS can hold crude or refined oil.

"Go green vote blue," urged Britain's Conservative Party on its way to victory over the incumbent Labour Party in 2010. Tory David Cameron's government has tied fighting climate change to ensuring national energy security. All major parties in the United Kingdom agree on energy security, cutting energy use and acting against climate change. At least on paper. Before becoming Labour prime minister in 2007, Gordon Brown noted that "every nation today is concerned about energy security."[1] His focus was national not continental energy security.

Why are Britons so attuned to national energy security? An eight-day petrol scarcity in 2000 is a major reason. The shortage caused panic buying, food scarcity, price spikes and serious disruptions to emergency and health care services. It showed the vulnerability of modern societies to an abrupt oil shut-off. Britons fared poorly, even though they use only 40 percent of the oil per person that Canadians do. Britain's petrol scarcity was politically caused. Its effects were similar to an external shut-off, but the government couldn't use companies' stocks as strategic petroleum reserves (SPRS) the way it would during an international oil scarcity. Like all other European Union countries, Britain must maintain SPRS. It has no government reserves, but orders companies to hold specified reserves to be used for public emergencies.[2]

In August 2000, fishermen blockaded English Channel ports in France to protest high fuel costs. Truck drivers, cabbies, ambulance drivers and farmers blockaded oil refineries to press for lower fuel taxes. With 14,000 of 17,000 gas stations closed, France's government caved and dropped gasoline taxes by 15 percent. France's concession ignited similar protests in Belgium, Germany, Ireland, Italy, the Netherlands, Poland, Spain and Britain. To cut use, Britain had raised the petrol duty by 3 percent above inflation each year since 1993. By 2000, Britons were paying the highest fuel taxes in Europe. Taxes and duties were three-quarters of the price. After seeing fuel taxes drop in France, where they were already lower, British farmers and truckers blockaded the Chunnel. Protests spread like wildfire, organized by CB radio and cell phones. Refineries were blockaded. People panicked.[3]

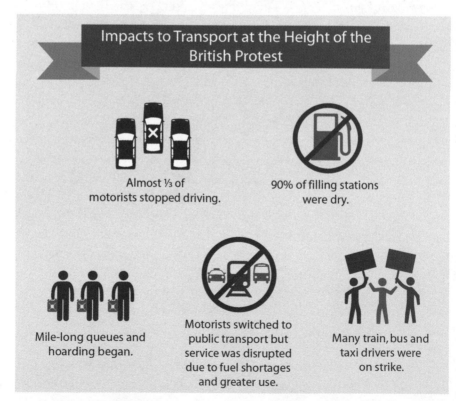

Impacts to Transport at the Height of the British Protest

Almost ⅓ of motorists stopped driving.

90% of filling stations were dry.

Mile-long queues and hoarding began.

Motorists switched to public transport but service was disrupted due to fuel shortages and greater use.

Many train, bus and taxi drivers were on strike.

Impacts at the Height of the British Protest

Transport:

- Almost a third of motorists stopped driving.
- 90 percent of filling stations were dry.[4]
- Mile-long queues and hoarding began.
- Many motorists switched to public transport, but service was disrupted because of fuel shortages and greater use. Many public transport drivers couldn't get to work.
- Train, bus and taxi services were reduced. Many cabbies were on strike.

Health care:

- Health emergency plans were put into effect. Services were rationed and emergency services restricted.[5]
- Staff, patients and supplies had trouble getting to health centres.
- Hospital food, medicines and blood were in short supply.
- Piecemeal local rationing tried conserving fuel for medical staff.[6]
- Some nurses were told to cycle to patient visits.

Food distribution:

- Fearing the worst, people stocked up on essentials.
- Some supermarkets rationed bread and milk.

Other:

- Some schools closed. There were too few teachers to "ensure pupil safety."
- Auto, steel, defence and aerospace industries were on the brink of reductions or shutdown, because of just-in-time shipments and interconnections.

Enter the Oil Companies

While popular anger ignited the protests, Shell and other petrol corporations defied Labour Prime Minister Tony Blair, refusing his demand to press charges against protesters. The contrast with usual corporate action against striking workers was notable. Even when police held back activists at most refineries, oil trucks failed to roll.[7] Oil companies supported protesters because they wanted to stop the Labour government restoring royalties on North Sea oil that the previous Conservative government had ended.

The *Guardian* newspaper explained: "When blockades began at Shell's Stanlow refinery in Cheshire, Shell could have had the protesters cleared in hours ... The oil companies fast recognized how they could use the situation .... At Britain's nine refineries some had significant protests. Others had fewer than twenty protesters, not even blocking the road.... It wasn't until [day five of the protests] that the government began to realize that the companies were deliberately holding back their drivers.... What happened during the last two days of the fuel crisis was a threat, a flexing of oil industry muscle.... The companies came close to showing they could bring down the government."[8] After winning, Big Oil let petrol flow again. In a secret deal, they delayed their victory so Blair's government could save face. Two months later, Labour cut the diesel fuel duty. Petrol companies immediately negated it by raising prices an equal amount.[9] Protesters and the government lost. Big Oil won.

Britain's near-miss showed the complacency that can beset petroleum-producing countries. Britain "lags behind other major European economies on gas storage because of its historic role" as a natural gas producer. It also showed the dangers of relying on energy imports and low strategic reserves. Britain deregulated energy in the early 1990s, hoping that the magic of the marketplace would provide reliable energy supplies. It didn't. Having no energy policy looked good until North Sea oil output began to decline, in 2000. Natural gas production started falling after 2005. Now Britain risks an electrical power shortfall, too. As *The Economist* magazine stated, "With gas too risky, coal too dirty, nuclear too slow and renewables too unreliable, Britain is in a bind."[10]

To gain energy security, Britain plans to lessen dependence on foreign oil and natural gas by cutting petrol use and beefing up emergency reserves and renewables. At 4.1 percent of Britain's total energy supply, renewables are small compared to Sweden's astounding 48 percent.[11] However, Britain plans a quick turnaround, with renewables producing 15 percent of electrical power by 2020. Britain obliges every licensed electricity supplier to buy a set and rising portion of energy from renewables.

The European Union requires all members to hold strategic petroleum reserves (SPRS) for three months' worth of normal use. Because North Sea oil made the UK a net oil exporter in the 1980s and 1990s, Britain's SPRS obligation to the EU was lower. In 2012, it had oil reserves for seventy-eight days. It's supposed to rise to ninety days. Few countries have natural gas reserves.

Britain's Trade Department argues that even if the probability of a natural gas shortfall is small, "the economic impact, if one does occur, could be many times as great as the cost" of the storage capacity. It recommends seventy-five days' worth of oil reserves. The UK's emergency natural gas stocks are not high by European standards. Canada should follow Europe's lead and hold natural gas reserves as well as oil reserves.[12]

Who Seeks Energy Security?

An international study of energy efficiency gains in the world's twelve largest economies put Britain at the top. Canada was second last, lagging behind even China.[13] Only Russia fared worse. Canada's usual excuses for energy overuse—huge unpopulated country and cold climate—were controlled in the study. Other large northern, sparsely peopled countries with high living standards, like Sweden, use up to 40 percent less oil per person than in Canada.[14] Despite having twice as many people as Canada, Britain uses less oil overall. Britain shows that Canadians (and Americans) can cut oil use in half without major lifestyle changes.

Taxes push Britain's petrol prices higher than in other parts of Europe and much higher than in Canada and the United States. Petrol prices, combined with a good mass transport system, dense cities and a culture of smaller cars, mean that Britons use much less gasoline than Americans and Canadians. Thirty percent of Brits don't own a car, compared to 10 percent of Americans. UK motorists drive half as far as Americans do annually. Overall, Britons use a fifth of the gasoline Americans do. Yet their living standards are comparable. Canadian road use is closer to the American level. But Canadians have 30 percent fewer cars and travel shorter distances—8,200 kilometres a year compared to 13,000 kilometres. Canadian road fuel use is 23 percent less.[15]

In 2008, to find out if Canada needs SPRs like the EU and the US, I called Matt Simmons in Houston in 2008. He headed Simmons International, one of the world's largest energy investment banks.[16] I was pleasantly surprised when Matt returned my call. "When I was working on Governor [George W.] Bush's campaign in Texas..." he began, making clear his political connections. Matt was an energy advisor to President Bush (2001–09). Not just a consummate energy insider, Simmons was a respected thinker on peak oil before he died in 2010. He spoke on topics like "Do Fossil Fuels Have a Sustainable Future?" and "The Oil and Gas System Is Sick"—not the usual oil boosterism fare. I asked Simmons if Canada needs strategic petroleum reserves.

"Do you Canadians use oil?" he replied. "Sure, Canada needs strategic oil reserves. The US exports about 1.5 million barrels of refined oil per day to Ontario and Mexico [in 2007–08]. If there was an international supply crisis, that oil would stop at the US border. You need national governments to deliver oil during emergencies to their own citizens. Governments often don't do a good job looking after their own people. They sure aren't going to take care of citizens in another country."

Matt told me about his second home in Maine. Until a rail strike and a cold snap in winter suspended propane deliveries by rail, he hadn't realized New Englanders depended so much on home heating oil supplies from Canada. The US has a small heating oil SPR in the region, used in 2012 after Hurricane Sandy, for such events. When the CBC's *The Current* did a segment on my 2008 report recommending SPRs for Canada, I proposed that they invite Simmons as a guest. Host Anna Maria Tremonti asked Simmons, "What would happen to shipments of oil to Ontario and Quebec in the event of an international supply crisis?"

Simmons: "We'd shut you off. We don't have an SPR big enough for us. The longer away you are from the source of the oil, the more vulnerable you are.... To think that our Strategic Petroleum Reserve in the United States is Canada's insurance policy is basically like one of my neighbours across the street saying 'You know, I've just realized I don't need fire insurance on my house because you have one on yours.' 'Yeah,' I'd say, 'I have it on my house.'"

"Would Canada be impacted if you suddenly had an oil shortage?" Matt continued. "Sure. It would be shut down. This is a facetious comment, but it illustrates the issue. If you moved all the people that now live in Quebec and Ontario to Alberta, then you'd probably be in good shape, but you wouldn't [export] any oil out of Alberta. It would all be used in Alberta. Your exports overwhelm the fact that you're a highly dependent importer."[17] Why are Americans in the know so conversant on national energy vulnerability issues, while similarly placed Canadians never talk about them? We need to. Canadians are more vulnerable to short-term oil supply crises than Americans.

SPRs for Short-term Crunches

Unlike most countries, Canada has the potential to secure oil for the medium term, but must reverse policy direction to do so. Later chapters focus on option one, a Canada-first energy and ecological security plan that combines energy independence with conservation to ensure that poorer Canadians have equitable access to domestic energy. Canada could withstand a longer-term, international oil decline by greatly cutting domestic use. Then Canada could meet international obligations on greenhouse gases, protect local habitats and cushion our transition to life after carbon fuels.

It surprises most Canadians to learn that we are at risk. It's counterintuitive. Alberta's Sands are deemed to hold the third-largest proven oil reserves in the world.[18] With all that oil, how could Canadians be vulnerable? Our insecurity results from exporting over two-thirds of our oil output, while Quebec's refineries have been importing over 90 percent of their oil, Atlantic Canada's about 83 percent and Ontario's 8 percent.[19] Oil imports to Quebec will undoubtedly fall when Enbridge Line 9 flows to Montreal, but even so, Quebecers will still likely depend on oil imports for more than half their oil. Eastern Quebecers and Atlantic Canadians will not be affected significantly by Line 9 and they will continue to rely overwhelmingly on imported oil. More US shale oil and refined oil is entering Ontario, so that province is not immune to oil supply crises either.

Canada is unprepared because current government leaders misperceive our eco-energy role and naively believe the market will always provide. Ottawa could easily end all oil imports and ensure that all Canadians get domestic oil, but it is too enthralled by Big Oil's export paradigm and political influence to do so. Along with Australia, Canada is the International Energy Agency member most vulnerable to short-term disruptions, because of a loophole in IEA requirements combined with complacent federal agencies like Natural Resources Canada and the National Energy Board. The IEA requires members to maintain emergency oil reserves (SPRS) and exempts only net oil exporters, assuming they will meet domestic needs before exporting surpluses.

Denmark and Norway are net oil-exporting IEA members and hold SPRS. Canada's foolish consent to NAFTA's energy proportionality rule means Canada must make available its current share of energy exports to the US, even if Canadians are freezing in the dark. The IEA should not exempt Canada from holding SPRS. Mexico rejected proportional, virtually mandatory exporting when it signed on to NAFTA, viewing it as a threat to their sovereignty and safety. Until Canada gets a Mexican-style exemption on proportional "sharing," Eastern Canadians will not have guaranteed access to conventional oil from Western Canada and Newfoundland. A future NDP or Liberal government could decide that domestic oil should go to Canadians before exports are allowed. That would be good, but could still leave unanswered the issue of shipping Sands oil via west-to-east oil pipelines.

The US Strategic Petroleum Reserve

Oil companies hold enough stock to ensure prompt delivery to customers and run their plants and vehicles between refills. Strategic oil reserves are different. National governments set them up to address emergencies that threaten the economy and their residents' well-being. The US has the world's biggest SPR, holding 727 million barrels. It can release 4.4 million barrels per day, enough to offset about 45 percent of imports. The US has used its Strategic Petroleum Reserve more than other countries. It has exchanged SPR oil with private oil companies eleven times, but has drawn down major amounts only three times: during the bombing of Libya in 2011, after hurricanes Katrina and Rita in 2005 and during the first Gulf War in 1991. These releases calmed world oil markets and showed that, contrary to free market ideology, US intervention successfully moderated oil price surges, discouraged speculation and brought benefits many times their cost.[20]

Most economists who oppose SPRs do so on the fundamentalist premise that markets should determine outcomes whatever the consequences. With the exception of Canada, that view has had little resonance among governments. American SPRs are a non-controversial, bipartisan issue, widely viewed as increasing energy independence and economic security and providing a cushion during short-term supply ruptures. The major US debate is on whether SPRs should also be used to hold down prices. The US stores most of its oil reserves in salt caverns on land near the Gulf of Mexico. The caverns are very cheap, with capital costs a fifth of those for above-ground tanks. Since salt caverns are 600 to 1,200 metres below the surface, geological pressure seals cracks. No oil seeps out. They are a concern, though. Seven barrels of fresh water are needed to dissolve the salt to make room for each barrel of oil.[21]

In 2000, the US set up a much smaller home heating oil reserve for northeastern states to shield the eight million US households that heat their homes with oil. By holding a million barrels of ultra-low-sulfate heating oil, Americans can buffer supply interruptions during severe winter weather.[22] In contrast, Canada has no home heating oil reserve for nearby Atlantic Canadians, half of whom use furnace oil for heat, and winters are even more severe there.

The US has a foreign oil storage program, through which Canada and other countries can place reserve oil in the US Strategic Petroleum Reserve, pay storage fees and supposedly get the oil when needed.[23] But rather than ship the oil directly to Canada, it could be sold in the US to offset oil diverted to Canada by tankers from nearby US sources. At first glance, it looks like

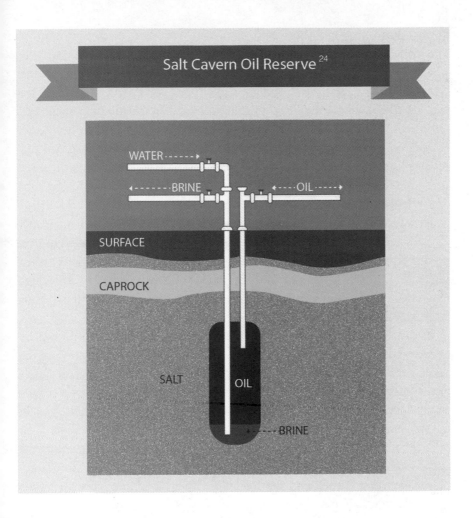

Salt Cavern Oil Reserve [24]

WATER

BRINE   OIL

SURFACE

CAPROCK

SALT   OIL

BRINE

an attractive option. But on reflection, relying on the US oil reserve is a bad idea. The US would gain too much leverage over Canada's eco-energy policies, allowing it to withhold or threaten to withhold Canadian reserve oil if we push independent energy, military or ecological policies. Matt Simmons warned Canadians not to expect to get American reserve oil in emergencies. His caution would apply even if we parked Canadian oil in the Gulf SPRS under agreement. Washington could apply *force majeure* in extraordinary events such as wars or acts of God to wriggle out of such an obligation. And it is only in such emergencies that SPRS are useful.

Protective Value of SPRs

The International Energy Agency was born after the 1973 Yom Kippur War, when Israel defeated Egypt and Syria. In retaliation, Arab OPEC members ended oil exports to the US and Holland for supplying weapons to Israel during the war. Big Oil tacitly helped cut oil output, as did non-Arab OPEC members. Oil prices quadrupled within three months everywhere, not only in the embargoed countries, greatly enriching both Big Oil and OPEC governments. As they fumed in gasoline lineups, Americans realized for the first time that their well-being depended on unfriendly countries. US leaders found such dependence intolerable and encouraged domestic exploration and energy conservation. The US focus on national energy security took hold. Washington initiated the International Energy Agency in 1974, an association of rich, oil-importing countries, to counter OPEC's oil boycotting power. The IEA had fifteen founding members, including Canada. Canada joined even though its role was and still is much more like OPEC countries—a net oil exporter. The IEA now has twenty-eight member countries.[25]

The IEA exhorts net oil-importing members to keep oil reserves equivalent to ninety days' worth of their net imports; implores all members to use demand restraint, fuel switching and surge oil output in international oil supply crises; and urges members to share oil with other member countries during severe supply disruptions.[26] If any IEA member suffers a 7 percent or greater fall in crude supplies, less-impacted members are supposed to share their oil. Each member is supposed to have a national emergency organization to coordinate with the IEA's emergency team. Each country decides how to cut oil use, boost oil output, and share oil with other members.[27] The IEA has no enforcement mechanism.

The US used to hold half of the world's strategic petroleum reserve capacity. No longer. Many other countries, especially China, are creating and filling their own SPRS.[28] The European Union requires its members to have SPRS. Nineteen of its twenty-eight member countries belong to the IEA.[29] IEA countries are obliged to have SPRS, except for net oil-exporters. Denmark and Norway have them anyway. Only Canada and Australia don't. Many non-IEA countries have SPRS, including China, Iceland, India, Iran, Israel, Mexico, Russia, Singapore, South Africa, Thailand and Turkey. That makes thirty-seven countries with confirmed reserves. Information on SPRS is spotty. Some countries are secretive about having SPRS, considering them a national security issue.

India created SPRS in 2004 because of concerns that oil imports, mainly from the volatile Middle East, met over 70 percent of its needs. China imports over half its oil and ambitiously plans to expand its sprs to one hundred days' worth of imported oil by 2020. Both countries also worry that rising oil imports will hurt their balance of payments and curb growth. Following France's 1925 lead, many countries require oil corporations to hold sufficient stocks to meet the country's needs, backed by penalties for non-compliance. Austria, Belgium, Greece, Italy, Luxembourg, New Zealand, Portugal, Sweden, Switzerland and Turkey have commercial-stock reserves, but no government reserves. Most countries with sprs are net oil importers, but a growing number of oil-exporting countries also have reserves: Denmark, Iran, Mexico, Norway, Russia and Saudi Arabia. Canada is again the odd country out.[30]

Three-Track Strategy for Building SPRs

If strategic petroleum reserves are a national insurance policy, how big should they be, what will they cost and where should they go? The target size appears to be clear: ninety days' supply of Canada's imported oil or 634,000 barrels per day in 2014.[31] That's 58 million barrels, or 8 percent of the size of the us Strategic Petroleum Reserve, a little less than one would expect for Canada in comparison with the us.[32] However, creating that large a Canadian spr could be a mistake. Some of Canada's current oil supply and demand can be altered through government action. Oil imported from the us by rail could be curtailed, and west-to-east oil pipelines could greatly reduce oil imports if ordered to ship more domestic conventional oil to eastern Canadians. Taking those actions could safely reduce the size of Canada's sprs. If so, they'd be less costly and fill faster.

We need a three-track strategy to quickly boost private companies' oil stocks for public emergency purposes, plan and fill public reserves, and reduce Canadians' oil use. sprs will take time to plan and be approved, built and filled. If everything goes smoothly, an unlikely event, it would be several years before Canada has a single barrel of government-held spr oil. That's too complacent a speed. To hurry things along, Canada could temporarily copy Europe and require oil companies to hold crude and refined oil in Eastern Canada for public use during oil shortages. Many petro-transnationals hold oil for such purposes elsewhere and see it as a cost of doing business. They will protest loudly but not be surprised at such a requirement.

If Canada follows this track, it can quickly enlarge reserves before government storage facilities are in place. Industry always has some oil stock on hand to ensure uninterrupted supplies for its customers. Between 2010 and

2012, refiners in Quebec averaged twenty-eight to thirty days of refined oil inventory, Ontario had twenty-two to twenty-six days, while Atlantic Canada had only fourteen to fifteen days.[33] These stocks are insufficient to ensure energy security. Refiners should be required to store ninety days' worth of oil to be used for public emergency purposes.

Doug Heath, Director of the Oil Sands and Energy Security Division at Natural Resources Canada and Canada's representative at the IEA, sees a shortage of refined product (diesel/gasoline) as a bigger danger than a shortage of crude oil. He cites unplanned refinery shutdowns that led to gasoline shortages in southern Ontario, Alberta and Nova Scotia. He thinks it reasonable to require oil companies to hold reserves of refined oil, though suspects they would complain.[34] While pursuing the first track of building up private oil reserves, Canada must simultaneously follow a second track of planning and doing environmental assessments on government-held SPRs. These steps will take time. Oil storage facilities can be expensive to build and slow to fill.[35]

The third track is to reduce oil imports. Every barrel of imported oil cut will reduce the size, cost and need for Canadian SPRs. Canada can reduce imports through conservation and by replacing them with Canadian oil. Deep conservation would enable Canada to jump from laggard to an international leader in reducing carbon emissions. However, even with a strong national plan, such as the one outlined in later chapters, Canadians cannot wean themselves off oil quickly. Transportation depends overwhelmingly on it. The fastest way to reduce or end oil imports is to replace them with domestic conventional oil.

SPRs: Getting Them in Canada

I've noted the reversal of Enbridge Line 9. If it carried Western Canadian oil almost exclusively, like it did until recently, the pipeline could replace up to 300,000 barrels of oil imports per day from the 650,000 barrels per day or so of net oil that Canada imports.[36] But Line 9 doesn't carry only Canadian oil anymore. Now it also brings a lot of US shale oil into Eastern Canada. TransCanada proposes to partly convert its natural gas mainline through northern Ontario into an oil pipeline to carry Western Canadian oil, including Sands and Bakken shale oil, to Quebec and then on to the Irving refinery via a newly built pipeline to New Brunswick. The Energy East oil line would be one of the world's most capacious oil pipelines, at 1.1 million barrels per day.

For years, I called for oil pipelines to bring Western Canadian oil to Eastern Canadians. Now that TransCanada and Enbridge are planning them, I should be pleased. If dedicated to serving Canadians with domestic conventional (non-fracked) oil, these pipelines could transform Canada from an oil-insecure country to head of the line. But both lines are about exporting Sands and other oil. They are not at all about providing Eastern Canadians with energy and environmental security. Big (mainly foreign) oil and major pipeline corporations seized on sending Sands oil east only after meeting obstacles to shipping it south through the Keystone XL line to the Gulf of Mexico and west via Enbridge's proposed Northern Gateway line and a twinned Kinder Morgan oil line to British Columbia's coast.

Whatever the motive for building or reversing the pipelines, wouldn't west-to-east oil pipelines make Eastern Canadians more oil-secure? Perhaps, but not necessarily. Instead of supplying domestic conventional oil to Eastern Canadians as part of a national eco-energy plan to transition Canadians off carbon fuels, both plans are Sands exporting ploys.[37] The incidental energy security benefit they could provide will end as soon as Big Oil finds that exporting Canadian oil is more lucrative than supplying Canadians with their own oil. The Harper government won't intervene to ensure that Canadians have energy security the way all other IEA countries do. "We don't tell the companies to put the pipelines here or there," Harper has declared.[38]

Despite major popular opposition, the National Energy Board approved the reversal of Enbridge Line 9. It's less certain that TransCanada's proposed Energy East oil line will be converted from natural gas and extended to St. John, New Brunswick. Opposition is mounting, mostly for environmental reasons. First Nations assert their sovereign right to decide what crosses their lands, and they worry about ecological calamities. Ontario and Quebec have seven conditions that must be met before they will approve the line. Although the provinces do not have the jurisdiction to block the pipeline, their strong disapproval would make it politically very difficult to proceed.

Could the Enbridge and TransCanada pipelines be transformed by a future progressive federal government into an energy and ecological security line for Canada? The answer is yes for a scaled-down Energy East line, but no for Enbridge Line 9. Canada cannot direct Enbridge to carry only Canadian oil in its pipelines through Great Lakes states. When the next international oil shortage hits, Washington may well declare *force majeure* and "national security" to halt Western Canadian oil's throughput to Ontario and use all the oil for itself. Reversing Line 9 cannot ensure energy and ecological security for Canadians.

The Energy East oil line is a more promising possibility, but only if it is repurposed from a Sands-exporting line to a Canada-first line bringing domestic, non-fracked conventional oil to Eastern Canadians for energy and ecological security reasons. To send domestic conventional oil to Ontario and Quebec, oil exports to the US will have to end. This would require Canada to challenge and overcome NAFTA's proportionality rule. In the next chapter, I discuss getting a Mexican-style exemption on proportionality, or if that fails, withdrawing from NAFTA after giving the required six months' notice. The outrageous thing about proportionality is that if the "market," NAFTA's euphemistic term for Big Oil, redirects all of Newfoundland's oil to Canadian consumers or carries only conventional oil on the Energy East line, it would be okay, even if the proportion of oil Canada exports to the US declines. But if citizens elect a government to direct oil companies to do these things, it would defy NAFTA.

If Energy East were transformed to carry only Western Canadian conventional oil, it would need drastic downsizing. Its proposed capacity of 1.1 million barrels per day is more than double the amount needed to replace all oil imports to Quebec and Ontario.[39] Canada's total conventional oil output is only about 1.4 million barrels per day, including Newfoundland oil.[40] Much of the conventional oil, which is slowly declining, will be needed in Western Canada and will not be available to put in a west-to-east pipeline. So Canadian conventional oil could not fill Energy East's proposed gargantuan volume.

Energy East's plan includes building a new oil pipeline from near Quebec City to Saint John, New Brunswick. This line would be useful for exporting Sands and US shale oil. It's not needed to bring oil security to Atlantic Canadians. TransCanada and the Irving group are planning an ice-free, deep-water port in Saint John to load the world's largest oil tankers with Western Canadian and shale oil to be carried to the globe's most lucrative markets. Why do Atlantic Canadians need to pipe in oil from far-off Alberta when Newfoundland's offshore oil fields produce enough to meet current demand in the four Atlantic provinces?[41] Much of Newfoundland's oil is now exported. It could be redirected to replace the 80 percent of imported oil that Atlantic Canadians now use.

Newfoundland's oil output is falling and will likely continue to do so despite expected gains from the Hebron field and possibly offshore Labrador. Atlantic Canada may have to use a small amount of Canadian conventional oil piped in from Quebec until Canada-wide conservation measures take hold and push the Atlantic region's oil use below its oil production level. Redirecting

Newfoundland oil will require some retooling of Atlantic Canada's refineries so they can refine Newfoundland's grade of oil and exclusively supply Atlantic Canadians. But retooling could happen a lot faster than planning, building, repurposing or reversing pipelines.

Scaled Down

The international standard for SPRS is to replace ninety days' worth of imports. While Canada can cut oil imports by conservation and by redirecting Newfoundland oil to Canadians, it should be building and filling public SPRS. How big should Canada's SPRS be? A federal government intent on setting them up should study projected oil imports given future west-to-east oil pipelines and oil imports by rail. To illustrate, let's assume that oil imports will fall to half of Canada's 2013 oil import level. Ninety days' supply to replace those imports in an international oil supply crisis is about *30 million* barrels. Hypothetically, if all of Canada's current oil output were poured into its SPRS starting on January 1, they would be filled by midnight on January 9.[42] Practically, it would take much longer, but it shows how little oil we are talking about. The US fills its reserves at 29,000 barrels per day. At that rate, Canada would fill its SPRS in just under three years.

Over the long term, we need to drastically reduce Canada's carbon emissions by greatly cutting Canada's oil use and production. Besides preventing the devastation of many local habitats, such a plan would end all oil imports. In addition to making Canadians more ecologically and energy secure, Canada could set its own oil price. Bargain gas prices lower than the world price, as arranged in the 1970s, encourage waste, but a made-in-Canada oil price could guard against the harmful booms and busts that extreme high and low oil prices bring.

A Canadian oil price that moves slowly around the world price averaged over the previous three years would bring on new supplies, enable producing provinces to collect maximum royalties and slowly discourage use. An oil price surge protection policy would be wildly popular among Canadian motorists, but could be open to abuses such as lowering gas prices before elections or subsidizing oil corporations when low prices hurt their profits. Accusations of both these abuses have been levelled against releases from the US Strategic Petroleum Reserve when there was no oil supply crisis.[43] Strict limits to price protection should be built into the terms of reference for Canadian SPRS, and an arms-length body could be set up to determine releases from it.

Siting Canada's SPRs

The siting and design of SPRs need research, public debate and hearings held in affected provinces. None of this has happened recently, but we can learn from work done in Canada in the mid-1970s and from the US. After the Arab oil boycott ended in March 1974, Canada received a study concluding that the idle Wabana iron mine, accessible by tanker at Bell Island, Newfoundland, could easily deliver emergency oil to Atlantic Canadians and Quebecers.[44] Wabana could and probably still can hold about 100 million barrels of oil, more than triple the 30 million that Canada needs today. Robert Stanfield, federal Progressive Conservative leader, demanded a "national policy to ensure that there is a minimum of ninety-day supply [of oil] held in storage at all times."[45] Considering the support and interest of the time, it's a pity the SPRs were never built.

If Canada were to realistically set up SPRs, there would probably need to be various sites. Dispersed storage facilities could enable quick distribution and overcome perceptions of regional unfairness and missed job opportunities. Politics requires that Quebec have its own SPR, likely sited on the St. Lawrence River and at least partly under Quebec control. SPR sites must be able to berth large oil tankers or receive conventional oil by pipeline from Western Canada. Facilities could be sited near refiners in Sarnia, Montreal, Quebec City and St. John or at oil-distribution access points. Reserves need to mainly stock crude oil, as refined oil degrades over time. Crude is cheaper to buy, store and move than refined oil. But some heating oil should be stored in Atlantic Canada, like the US does in New England.

Should Canada use the sixty-one salt caverns, located about six hundred metres below ground, in Lambton County, near Sarnia? They've long stored hydrocarbons. Their capacity, rated at about 31 million barrels of oil, is enough to hold all of Canada's SPRs.[46] A drawback would be a one-time use of fresh water to dissolve the salt. Lambton is well placed for SPRs. Three of Ontario's refineries are in nearby Sarnia, and the fourth is at Nanticoke, connected to Sarnia by pipeline.[47] Their location, though, confines their use to Ontario, as shipping from there by rail is costly and dangerous.

Cost of SPRs

Buy low and sell high. Bill Richardson, Energy Secretary to President Bill Clinton, advocated this ancient merchants' motto for the US Strategic Petroleum Reserve. Matt Simmons called SPRs a "federal reserve bank of oil." In 2015, international oil prices were low, a perfect time to buy SPR oil at low prices and watch the value rise. China has been doing precisely this at a fast pace. Whatever the cost of setting up SPRS, other countries conclude that it pales in comparison to the economic and human costs of not having them. Both the havoc caused in the US by the 1973–74 oil boycott and Britain's oil strike in 2000 were foretastes of facing oil supply crises without SPRS. Charles Hendry, Britain's Conservative energy minister, declared in 2011 that "there is a price for energy security but it's nothing like as high as the price of energy insecurity."[48]

Natural Resources Canada has not studied current costs of Canadian SPRS, but we can extrapolate from the American case. Buying oil to fill the SPRS is by far the biggest cost. But governments can recoup it and even make money when it sells oil at higher prices during oil emergencies. Building storage units is the main permanent cost. Above-ground storage in the US goes at $15 to $18 a barrel. Salt caverns are much cheaper: about $3.50 a barrel.[49] The 1974 Canadian report rated the building cost of storage at Wabana at an astonishingly low eighteen to twenty-seven *cents* a barrel, 4 percent of the construction costs then of above-ground steel tanks. Converting those numbers to today's prices, Wabana could cost seventy-five cents to a dollar a barrel, far below the current $3.50 for the US salt caverns.

If Canada's SPRS held 30 million barrels, above-ground storage could cost around $500 million, salt caverns about $100 million and Wabana around $30 million. These are ballpark figures; a new study is needed to determine actual costs. Canada could get SPRS fairly cheaply using Wabana, Lambton county's salt caverns and probably a more costly Quebec site. Even if costs are at the high end, $500 million is not a great price to pay for Canadians' oil security and to curb wild gasoline price spikes.

Economists generally cite filling costs as pure expense.[50] They would never commit this elementary error when looking at corporate ledgers. A store of oil would be deemed an asset to be balanced against costs. Governments generally fill SPRS when oil is plentiful and cheap, and sell during supply crises when prices surge. The average filling cost for the US Strategic Petroleum Reserve was $29.08 a barrel between 1976 and 2008. Filling Canada's SPRS would cost about $1.5 billion at $50 a barrel, $3 billion at $100 a barrel, and $4.5 billion at $150 a barrel. These major but affordable costs are not permanent expenses. From 1976 to 2009, sales from the US reserve produced *net* revenues of $22 billion to the US Treasury.[51] While curbing gasoline price surges, they provided Americans with energy security. SPRS could earn Ottawa net revenue.

No Studies on Canadian Energy Security

I wrote to the National Energy Board (NEB) in 2007 to get its latest report on Canada's energy contingency plans. The NEB was set up in 1959 to promote safety and energy security in the Canadian public interest and do studies around its mandate.[52] Environmental protection and efficient energy infrastructure were later added. I was astonished when the "NEB Communications Team" wrote to me that "unfortunately, the NEB has not undertaken any studies on security of supply."[53] What? That's their mandate. If not the NEB, then who is studying energy security for Canadians?

No one, it seems. Not Natural Resources Canada either. Nor has there been a government study on Canadian energy security since my inquiry. There will be no study until the Minister of Natural Resources requests it. No one expects such a request while Harper is in charge.[54] I also asked the NEB if Canada is considering setting up SPRs like the other twenty-seven members of the International Energy Agency have done. Canada was "exempted from establishing a reserve," the NEB team wrote, "because Canada is a net exporting country whereas the other members are net importers."

The exemption is true but misleading. Drafted in the 1970s, when national energy self-sufficiency was widely lauded, including by federal Progressive Conservative leader Robert Stanfield, the IEA assumed that net oil exporters would supply their own people before exporting surpluses. On this premise, such countries don't need strategic reserves. But since the advent of neoliberalism, countries don't act sensibly, least of all Canada. How sensible is it to export over 70 percent of the oil we produce while importing 40 percent of the oil we use? Those import percentages are bound to drop after Enbridge Line 9 is re-reversed and may drop further if and when the Energy East oil line is completed. Even so, Canada is leaving its people vulnerable by abandoning energy security to the whims of Big Oil and by still acting like an oil-importing country.

Canada is the thirty-seventh most populous country in the world but the fifteenth greatest oil importer. The NEB is wrong that all other IEA members are net oil-importers. IEA members Denmark and Norway are net exporters too. Denmark has no IEA obligation to hold SPRs, but must hold them as a European Union member. Denmark imports as well as exports oil, so it protects its people by requiring industry to hold eighty-one days' worth of oil for national use. Norway produces much more oil than it uses, but prudently provides for its people by requiring oil companies to hold twenty days' worth of stock for the domestic market. Norway is not a European Union member, but does this for security reasons.[55]

The IEA is not as confident as Ottawa about Canadians' invulnerability. "Canada is not immune to the risks of supply disruption," the IEA flatly states. When an international oil supply crisis hits, Canada's only recourse is to declare an emergency and require oil companies to draw down their commercial stocks. Provinces, not the federal government, hold the authority to force demand-restraint measures.[56] That is like waiting until the dykes broke after Hurricane Katrina hit New Orleans rather than strengthening them beforehand. The NEB's explanation for Canada's exemption on SPRS doesn't add up. In a country where winter is the dominant season, Canada needs an energy security plan.

The Politics of Indifference

If the case for holding oil reserves is so compelling, why have Conservative and Liberal governments failed to create them as Canada's traditional allies have done? It's partly because of the stories we tell each other about Canada having so few people in such a large land of unlimited resources. The idea of oil shortages here seems unreal. Alberta's Sands are deemed to have enough recoverable oil to power Canada at current rates of consumption for over two centuries if earmarked for domestic use only. But that's way too much carbon to release into the atmosphere.

We also share a naive faith peculiar to this continent that undirected markets will provide for people's needs. The 2008–09 Great Recession shook many people's faith that markets know best, but many have not transferred that skepticism to oil supplies. But it's more than naïveté about limitless resources and markets. The federal Conservatives, Alberta's Conservative governments and, above all, Big Oil have pushed a continental energy agenda since Canada's National Energy Program ended in the early 1980s. Once they backed down from opposing NAFTA's virtually mandatory energy-exporting rule in 1993, the federal Liberal party toed the same continental energy-sharing line.

Ottawa has steadfastly committed to using Canadian energy exports to help ensure "North American" energy security, hiding the fact that North American really just means American. How else can we explain why, almost alone, Canada has no energy security and independence strategy? Most Canadians would balk if they knew they don't have first right to their own energy resources. But using the language of energy security could be the undoing of the federal Conservatives and the petro-elites. How can they promise energy security to Americans and not Canadians? Canadians need to be aware that no one is looking after them before they push Canada's governing elites to put Canadians first in line to get their own energy resources.

If the US can have a NEP and officially pursue energy security, self-sufficiency, energy independence and domestic ownership and control of energy companies, how can it be anti-American to demand the same for Canada? Energy security for Canadians is both utterly necessary and revolutionary. Achieving it will mean overturning the petro-elites' power. The business-as-usual crowd who run this country do not want Canadians to realize they can take back their energy and environmental future.

Chapter 4
NAFTA and Proportionality: A Devil's Bargain

We will not weaken or renegotiate any energy provisions of the FTA or the NAFTA. Specifically, we will not allow the Canadians to opt out of the proportionality clause or to limit its coverage solely to oil.

—US President Bill Clinton, 1993

[Prime Minister Chrétien] came back with zero, zilch, nothing.... It's a complete sellout.

—Jean Charest, Progressive Conservative MP

As the age of easy oil passes, Canada is one of the few industrial countries with enough conventional oil to last decades. We could shut down Alberta's Sands entirely and still have enough oil to get Canadians through the transition to a low-carbon future. However, NAFTA's proportional energy-sharing clause stands in the way. Canada squanders its advantage by virtually giving the United States first access to over two-thirds of all its oil, including the less-dirty conventional kind. Simultaneously, Canada imports a lot of oil. It was about 40 percent in 2014, but will likely fall somewhat in the next few years, though still not enough to limit exposure for many Eastern Canadians. Thus, despite our rich resource endowment, Canada can't guarantee that in an international oil supply crisis Eastern Canadians will get first call on domestic oil.

To ensure oil security for Canadians and to combat climate change at the same time, Ottawa could order TransCanada's proposed Energy East pipeline to fully supply Ontarians and Quebecers with domestic conventional (non-fracked) oil. It could also redirect all Newfoundland oil to Atlantic Canadians. Both measures would end all oil imports. Canadians would also have to markedly cut their oil use. Until those measures are in place, Canada could create and fill strategic petroleum reserves. But NAFTA's proportionality rule undermines Canada's ability to ensure that residents get priority access to their own carbon energy, curbing emissions and protecting habitats.

The government of no other developed country is forbidden from guaranteeing its citizens first access to their country's energy resources. This matters because international oil supplies will almost certainly be disrupted in the near future. The only questions are when, how long shortages will last and the extent of the shortfalls. Canadians are expected to have faith that the market will provide. But the oil market is dominated by a few dozen petro-corporations that don't care if Canadians get first access to their own oil or natural gas; their only mission is to sell to the highest bidder. NAFTA was written before 9/11 and the "security trumps trade" era. Using the same logic for Canadians, ecological and energy security should trump NAFTA. Only Ottawa can decide that energy and ecological security for Canadians overrides the market and NAFTA.

If Canadians were its only consumers, Alberta's Sands could supply us for about three hundred years.[1] Michael A. Levi, in a report on Alberta's oil sands for the New York–based Council on Foreign Relations, explains the impossibility of reconciling rising Sands production with falling carbon emissions. He asks readers to "Imagine...that oil sands emissions rose as expected over the next two decades and then stabilized in 2030 while total US and Canadian emissions dropped by 80% by 2050.... Oil sands' emissions then become equivalent to about 10% of US emissions by 2050, representing almost all emissions from Canada at that point."[2]

Canada cannot shut down all other uses of oil just so Alberta's Sands can reach whatever output level Big Oil wishes. But Canada can meet its target of reducing carbon emissions by 80 percent if it phases out Sands oil and relies instead on our slowly falling output of conventional oil and natural gas as transition carbon fuels to get Canadians to a low-carbon future run on renewables. Canadians cannot be convinced to seriously conserve if they don't see a direct link between their actions and their country's emissions. Unfortunately, NAFTA's proportionality rule means the more oil we conserve, the more Big Oil will export whatever is saved. If the conserved oil is sent to the US, Canada's export obligation under NAFTA's proportionality rule will grow.

What Is Proportionality?

Proportionality is "unique in all of the world's treaties," writes Richard Heinberg, a noted California energy expert. Cyndee Todgham Cherniak, a Toronto trade lawyer, says the energy chapter is unique for a trade agreement.[3] There are only three free trade agreements in the world that have energy chapters, and the other two don't have NAFTA-like proportionality clauses. It's unclear how many other countries the US has tried to impose an energy proportional sharing chapter on, but it is clear none has bitten. Heinberg concludes that "Canada has every reason to repudiate the proportionality clause, and to do so unilaterally and immediately."[4]

Proportional sharing requires NAFTA members to make available the current share of energy exports to other member countries even when facing energy shortages at home. Strictly speaking, "making energy available" is not a mandatory exporting obligation—but it might as well be. Canadians could theoretically buy up all the Canadian oil or natural gas made available for export, but buyers of crude oil and wholesale natural gas are not ordinary Canadians or small businesses. They are huge petro-corporations, most of them foreign owned and controlled, operating on a profit basis. Their buying decisions have nothing to do with supplying Canadians with our own energy. Unlike most major oil- and natural gas–producing countries, Canada has no government-owned energy firm to protect its citizens' energy and ecological interests.

The proportionality clause says that if the government of any NAFTA member country takes action that cuts the availability of energy for export to another NAFTA member country, it must continue to export the same proportion of total "supply" that it has over the previous three years.[5] If it cuts energy available for export to another member country, it must also cut the supply of that energy domestically to the same extent. When NAFTA came into force in 1994, it built upon and superseded the 1989 Canada–US Free Trade Agreement (FTA). Mexico was added and the agreement altered, but the FTA's energy proportionality clause was retained.

The 1980s neoliberal thinking on energy that underlay the Canada–US FTA was this: it's good if a corporation decides to supply Market A rather than Market B, for profit reasons, but it's bad if governments protect residents by ensuring that Market A is served for energy security, energy sovereignty or ecological reasons. Energy exports can rise or fall through changes made by the "market," but elected officials cannot intervene to provide energy security to citizens who elect them. If TransCanada, for example, decided to ship more Western Canadian conventional oil to Eastern Canadians, it would not violate NAFTA's proportionality rule. But if Ottawa ordered TransCanada to send Eastern Canadians conventional oil for energy and ecological security reasons, it would likely contravene NAFTA. The clause casts a chill over debates on Canadian energy and ecological options, but has never been invoked. Since the FTA and NAFTA began, no Canadian government has been inclined to put Canadian environmental and energy interests first.

In addition to proportionality, NAFTA's article 605 throws another curveball. Exporters can't disrupt "normal channels of supply" or "normal proportions among specific energy" goods by, for example, substituting light crude for a heavier variety. Proportionality is based on total "supply," not "production."

The distinction matters and shows proportionality's bizarre logic. Supply includes domestic output, drawdowns of domestic inventory and *imports*. The almost 700,000 barrels of oil a day that Canada imports are added to domestic oil to form Canadian "supply." In 2011, Canada had to offer 73 percent of its total domestic oil output to the US.[6]

The gap between total supply and domestic output is much less for natural gas because we import only 15 percent of total supply, though such imports are rising.[7] When imports are subtracted from Canada's natural gas exports, Canada sends just under half of its net natural gas output to the US. In future, it will be easier to end liquefied natural gas exports to Asia than to stop natural gas exports to the US because it would not violate a treaty.

Mexico Exempt from Proportionality

When NAFTA started in 1994, Mexico was America's third-greatest supplier of foreign crude oil. It still is. Canada was second then but has since zoomed past Saudi Arabia to become America's greatest foreign supplier by far. During NAFTA negotiations, Mexico resisted strong US pressure to sign on to proportionality. Consequently, Mexico is a full member of NAFTA but is exempt from the energy proportionality rules. With an eye on domestic opinion, Mexico issued "five Nos." One "No" was a rejection of proportional sharing. Formally guaranteeing oil exports to the U.S. contravened Article 27 of Mexico's Constitution: "Under no circumstances may foreigners acquire direct ownership of lands or waters."[8] Sovereignty over Mexico's energy resources was and still is a revered part of Mexico's heritage and identity. Acceding to proportionality would have violated one of the proudest chapters in Mexican history. Every March 18, Mexicans celebrate Energy Independence Day to mark the day in 1938 when Mexico expropriated the foreign-controlled oil companies.

Publicly, Mexican governments portray themselves as the nation's resolute defenders. Privately, they were opening up Pemex, the popular, nationalized oil company set up in 1938 to replace big foreign oil. Fabio Barbosa, an energy expert at UNAM university in Mexico City, wrote me to say that "at the start of the nineties," when Mexico was loudly rejecting proportionality, it "was exporting everything it could."[9] In 1990, Teresa Gutiérrez-Haces, a political economy professor at UNAM, invited Canadian critics of the FTA to Mexico City. It was the first time Mexicans had heard about energy proportionality. Gutiérrez-Haces and others set up RMALC, the Mexican network on free trade, to stop Mexico joining NAFTA.[10] She contends that the Canadian critics, who had just lost the FTA battle in Canada, alerted Mexican activists and government to proportionality's dangers.

Did Mexico cede anything to get out of proportionality? Yes. It had to allow bids from US and Canadian corporations for government procurement contracts, a departure from allowing only domestic bids. Mexico also had to open parts of its petrochemical and electrical industries to foreign ownership.[11] However bad it was for Mexico, debt agreements were temporary deals that lacked the greater permanence of proportionality for Canada. Canadian energy researchers Terisa Turner and Diana Gibson write that "though Mexico was forced to agree to some compromises, it maintains control over the three key aspects to energy resource sovereignty: pricing, production and export levels, confirming that Canada could choose a different path."[12]

America's Version of Proportionality

Unlike Mexico, the US has no formal exemption on proportional energy sharing. Until recently, it may as well have. But now the US sends Canada about half the oil it imports and growing amounts of refined oil, including gasoline. However, since the US is and will remain a net importer of a quarter to a third of its crude oil, it in effect re-exports to Canada the oil it imports.[13] NAFTA's proportionality rule therefore does not really apply to the US.

It has not really applied to the US regarding natural gas either. But that is rapidly changing. Fracking in the Barnett shale play near the US Gulf coast and the Marcellus and Utica shale plays in New York, Pennsylvania, Ohio and West Virginia are spectacularly boosting US natural gas output and displacing natural gas imports from Canada. The US is projected to become a natural gas net-exporter by 2019.[14] Gas exports to Ontario and Quebec from nearby Pennsylvania and New York are becoming significant. This means that NAFTA's proportionality rule could in future apply to significant quantities of US natural gas exports to Canada. This obligation could hinder even US energy security when American natural gas output starts to fall. By 2007, when the US reached its peak of importing natural gas, Canada supplied 83 percent of it.[15] US dependency on natural gas imports at their height— 15 percent—never matched their import level of oil, 60 percent.

When the Canada-US Free Trade Agreement was negotiated in the late 1980s, US natural gas production had been declining for fifteen years, and it was not expected to recover. Getting unlimited access to Canada's natural gas through proportional sharing was one of Washington's major goals in the FTA. To do so, it needed to end Alberta's and Canada's long-standing protections of natural gas for domestic consumers. In 1949, Alberta's Social

Credit government had passed legislation that assured Albertans' first right to their own natural gas and placed other Canadians' needs above those of exports. Ten years later, John Diefenbaker's Progressive Conservative government in Ottawa copied Alberta's special protection of natural gas. When his government set up the National Energy Board in 1959, it reserved twenty-five years of "proven" supply of natural gas for Canadians, before it would issue long-term export permits. These Alberta-first and Canada-first policies were in place until the late 1980s, when the Canada–US FTA was being negotiated.[16]

Greenhouse gas emissions are growing faster from producing Sands oil than from Canadians using oil. Does NAFTA's proportionality clause allow Canada to cut oil exports to the US for conservation reasons, or because the

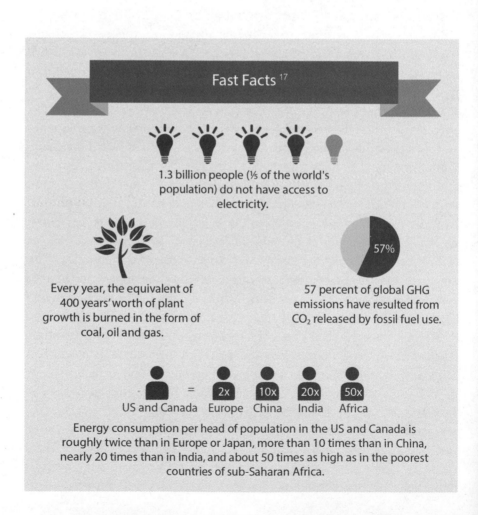

Fast Facts [17]

1.3 billion people (⅕ of the world's population) do not have access to electricity.

Every year, the equivalent of 400 years' worth of plant growth is burned in the form of coal, oil and gas.

57%

57 percent of global GHG emissions have resulted from CO_2 released by fossil fuel use.

■ = 2x   10x   20x   50x
US and Canada   Europe   China   India   Africa

Energy consumption per head of population in the US and Canada is roughly twice than in Europe or Japan, more than 10 times than in China, nearly 20 times than in India, and about 50 times as high as in the poorest countries of sub-Saharan Africa.

oil is dirty? No. The only grounds for reducing the oil available for export to the us is to be running out of it. Unfortunately, the Sands' abundant oil won't let us do that. Canada is allowed to cut oil output under proportionality, but Canadians would suffer. If Canada cut oil output by 10 percent, it would mean a twenty-seven-day oil shortage in Canada. If, on the other hand, we adopt a Canada-first oil policy by exiting from NAFTA's proportionality rule and phasing out oil exports, we can cut oil output by over two-thirds without shuttering any gas stations in this country. While proportionality prohibits Canada from reducing energy exports for conservation reasons, the GATT (the General Agreement on Tariffs and Trade) allows it. Canada would revert to GATT rules if it ends proportionality or leaves NAFTA.[18]

Long-time US Attempts to Grab Canadian Energy Resources
Washington has aimed to gain control over Canadian energy resources since 1952, when the Paley Report identified Canada as the most secure source for many of the kinds of raw materials their military economy needed.[19] In 1970, three years before Arab countries in OPEC embargoed oil exports to the us, Labor Secretary George P. Shultz warned that oil-exporting countries were banding together. He recommended that the us seek "safe" supplies from Canada.[20] Joe Greene, Canada's energy minister in Pierre Trudeau's government, was eager to give the us access to Canada's energy resources and criticized us quotas that limited oil imports from Canada. Schultz's and Greene's wishes came true two decades later in the 1989 Canada–us Free Trade Agreement (FTA).

The FTA and NAFTA are not the only ways for the us to secure control over Canada's oil and natural gas resources. us corporate ownership is also very effective. The us has used American-owned oil corporations as extensions of its foreign policy since before World War II.[21] us FTA negotiators insisted on "proportional sharing" to prevent Canada from reducing oil exports to the us, as it had after the 1973–74 Arab oil embargo.[22] To ward off the genuine threat of oil shortages in Quebec and the Atlantic provinces, Canada had redirected oil it had been exporting to the us to those provinces instead.

For us negotiators and their corporate backers, getting proportionality was the major coup of the 1989 FTA. It hadn't occurred to the us National Association of Manufacturers (NAM) that "free trade" could include a guaranteed right to Canadian energy. But R.K. Morris, NAM's director for international trade, was delighted when it did. "When we got such a great deal on energy [in the FTA]," he told journalist Linda McQuaig, "We were crusaders for the deal."[23]

Leaving NAFTA

Exiting from NAFTA would not automatically release Canada from its proportionality obligations. Canada would also have to withdraw from the earlier FTA with the US. An Act of Parliament, to participate in an agreement, can be changed more easily than changing the international agreement itself. The real issue is political and not narrowly legal. An informed Canadian public could put pressure on politicians who appear to amend or even dispense with NAFTA while preserving proportionality. The US got much better access to foreign oil through the FTA and NAFTA than it did by starting the International Energy Agency after the 1973–74 Arab oil embargo. US Secretary of State Henry Kissinger persuaded other industrial countries to sign a non-binding oil-sharing pact to counter OPEC's oil cartel power, obliging IEA members to share oil if international supplies fall 7 percent or more.[24]

Although it is a net oil exporter, Canada has been an IEA member from the start. This means Canada is obliged to share oil in a severe shortage with fellow members, most of whom are net-consuming countries. It's a bit like being the chicken in the middle of a wolf pack. Today's IEA is stuck with an emergency response system set up to meet 1970s conditions. Western governments assumed then that oil supply disruptions would be short-lived, oil-producing countries would always want to sell their oil, and the West could remove any government that held it to ransom by bombing, invading or seizing its oil supplies.[25] All these assumptions have been disproven. The peaking of deliverable conventional oil in the world shows that supply disruptions are likely to be long-term. The 2003 invasion of Iraq actually reduced its oil output for four years.

The IEA's sharing system has never been tested fully. When Hurricane Katrina devastated New Orleans in 2005, international oil supplies fell by only 1.5 percent, well below the IEA's 7 percent trigger level. Out of goodwill, Canada and European IEA members voluntarily sent the US additional oil for sixty days. But no general sharing among IEA members occurred. Unlike trade agreements, where failure to comply can spark trade sanctions, the IEA has no enforcement mechanism. From the US perspective, leaving natural gas out of the IEA agreement has been a shortcoming. In contrast, NAFTA covers all forms of energy, including natural gas, and it gives the US continuous rather than just emergency access to Canadian oil.

Proportionality is a lasting way to get someone else's oil. It doesn't take an extraordinarily deep shortage to trigger it. In contrast to its long-standing

Middle East policy, the US gets first access to foreign oil, Canada's, without costly invasions or military bases. However, because of its "addiction to oil," the US is still vulnerable to politically motivated oil embargoes. If Americans reduced consumption by a third they could live entirely off their domestic oil output and still use more than twice the world's per capita level.

America's National Energy Policy

After taking office in the hanging-chads election of 2000, President George W. Bush appointed Vice-President Dick Cheney to develop a National Energy Policy (NEP). Cheney's group recommended energy security, energy self-sufficiency and support for American energy firms. The latter was soon demonstrated when mounting US pressure persuaded CNOOC, a Chinese state oil company, to end its bid to take over Unocal, an American oil company.[26] It was almost as if Cheney had copied Pierre Trudeau's NEP, which had so incensed Ronald Reagan and US oil corporations.

Mexico's assertion of energy sovereignty appeared to pay off, at least rhetorically. Cheney's national energy policy report expressed respect for Mexican sovereignty, in contrast to Canada's, where it assumed the US could take as much Canadian energy as it wanted. "Canada's deregulated energy sector," the report stated, "has become America's largest overall energy trading partner, and our leading foreign supplier of natural gas, oil and electricity." It noted that the continued development of Canada's recoverable heavy oil reserves "can be a pillar of sustainable North American energy and economic security."[27] It mattered that Mexico asserted energy sovereignty during NAFTA negotiations and Canada did not. It matters even more now, in the current era of rising climate change disasters, the end of easy oil and energy insecurity.

The US promotes diversifying oil imports, so it is not too reliant on any one country. While the US expects to import much more Alberta Sands oil, it does not anticipate that most imported oil will come from Canada. Richard N. Haass, President of the Council on Foreign Relations, wrote that the Sands are not "critical to US energy security." Better security is the benefit the US gets by getting Canadian oil. Even if production costs are generally higher in Canada, "externality costs" associated with US military and diplomatic operations are nil. There's another advantage to Canadian oil for US petro-corporations. Despite costly production, Sands oil has been a bargain for US refiners since 1997. An oil glut in mid-continent due to bottlenecks in pipeline capacity has often led to a much lower oil price than for offshore oil imports.[28]

Origins of the FTA's Proportionality Obligation

Canada entered the FTA and NAFTA under very different conditions from today's. Then it was widely assumed that the world had plenty of cheap, easy oil. Few had heard of climate change chaos. With Canada's huge land mass, small population and seemingly endless resources, Canadians were especially susceptible to these assumptions. Before 9/11, securing energy supplies was on few Canadians' minds. It made sense to many that Canada should seek virtually guaranteed access to US oil and natural gas markets that in the past had been partly blocked. In return, Canada would give the US first call on the majority of our apparently limitless energy reserves. None of these assumptions now hold.

When Obama stopped TransCanada's Keystone XL oil pipeline in January 2012, he broke the understanding underlying the proportionality rule by ending unlimited US access to Canadian energy. NAFTA was broken, but it is not yet widely acknowledged as such. In other countries, governments are grappling with carbon emissions and energy security. But the business-as-usual crowd in Canada expects to carry on as before. We can't. Canada has only a decade of proven supplies of conventional oil and natural gas. Yet Canadian governments keep pushing forward, unaware that we're heading for a cliff as dangerous as Alberta's Head-Smashed-In Buffalo Jump.

Albertans are in for a shock. They assume their province has decades of natural gas remaining, yet it has only nine years of "estimated established reserves." Each year there are new finds, so Alberta will not run out in nine years—but the trend is strongly downward. In the past twenty years, the province has added new net additions of natural gas in only one year, 2008. Yet instead of conserving natural gas so Albertans can use it to heat their homes, schools and workplaces during the province's frigid winters, Alberta exports over 40 percent of it and uses more and more to produce Sands oil. The province used to sensibly give priority to supplying Albertans first and other Canadians second, before any natural gas could be exported.[29]

Albertans saw the province's natural gas as their birthright in a way that oil never has been. In 1949, the province reserved the supply of natural gas for Albertans for thirty years and required licences to remove gas from the province. These rules remained until the late 1980s, when the province dropped protection of natural gas for Albertans to fifteen years and ended the distinction between exports and natural gas for Canadians outside Alberta. By the late 1990s, Alberta no longer had enough established reserves of natural gas for all Alberta users, so it introduced the slippery language of protecting "core" Alberta users for fifteen years.[30] Alberta's Conservative government finally dropped the pretence of reserving natural gas for Albertans in 2013, when it replaced the Energy Resource Conservation Board with the Alberta Energy Regulator.

Alberta's former government put an optimistic spin on future supplies of the province's natural gas because of a "produce and profit" today mentality: let the future take care of itself. But it won't. The Alberta Energy Regulator estimates that Alberta has only 33.5 trillion cubic feet of remaining established reserves of natural gas.[31] Since the province produces about 3.6 million trillion cubic feet a year, those reserves won't last long. And neither coal-bed methane nor shale gas will save the day for Alberta. After over a decade of development, coal-bed methane produces only 8 percent of Alberta's natural gas output, while shale gas produces a miniscule 0.1 percent.[32]

A growing share of Alberta's dwindling natural gas is used to make Sands oil. Ziff Energy predicts that such gas use will almost triple between 2011 and 2020 and equal the natural gas Albertans used in 2011.[33] Natural gas use to make Sands oil already burns 10 percent of all the natural gas used in Canada.[34] CAPP and Statistics Canada do not tell us what percentage of Sands oil is exported, but it's likely about the same as Canada's total oil exports—over 70 percent. So a growing portion of Alberta's falling supplies of natural gas is embedded, in effect, in Canada's growing export of Sands oil. Under present production techniques, the rise of Sands oil output depends on greater and greater cannibalization of Alberta's natural gas.

On the precautionary principle, Alberta should return to the rule that it will not permit the removal of any natural gas from the province unless there are at least fifteen years of "established" supply for all Alberta users. If there are natural gas surpluses beyond Albertans' long-term needs, they should be sold to other Canadians. To stretch out established natural-gas reserves for Albertans, the province will have to quickly phase out gas exports and its rising use to fuel the Sands. If Alberta ends natural gas exports to the US—1.52 trillion cubic feet in 2011—Alberta's established natural gas reserves would extend from nine to about fifteen years.[35] Phasing out natural gas use in the Sands would lengthen Alberta's proven natural gas reserves still further. Gaining a decade of natural gas for Albertans is crucial to Alberta's transition to a low-carbon society.

China is also now taking climate change seriously. The Europeans have been there for some time. Canada will not be able to resist international pressure to take serious action to cut greenhouse gases for long, meaning expanding the

Sands will become a non-starter. So although Canada has lots of Sands oil, we will be unable to use most of it. That leaves Canada with two options: we could rely increasingly on oil imports from unstable OPEC countries or stretch out the lifespan of our conventional oil by phasing out exports and seriously cutting domestic use.

Proportionality: The "Alberta Chapter"

Marci McDonald, Washington bureau chief for *Maclean's* magazine, covered the US capital during Brian Mulroney's prime ministership (1984–93). In her book *Yankee Doodle Dandy*, she contends that James Baker, Ronald Reagan's Secretary of the Treasury, got the energy proportionality provisions on the table during a dramatic showdown at the FTA talks. Baker was a prominent oil industry lawyer based in Houston. McDonald contends that the six giant US oil corporations lobbied for unrestricted access to Canada's energy.[36] We don't know how often they bent the Reagan government's ear on Canadian oil and gas in the FTA talks, but if business was conducted as it was several years later in the NAFTA talks, it was many times a day. US officials held an astonishing 12,000–16,000 meetings with corporations during the NAFTA talks. Mexican and Canadian negotiators often complained that their US counterparts acted like agents of the private sector.[37] Was McDonald right that it was US oil corporations that insisted on the virtually mandatory proportionality clause? Insiders know who dreamed up and pushed the FTA's obligatory version of proportionality, but they aren't talking. Canadian authorities were determined that it stay that way. Pat Carney, Canada's Energy Minister during the FTA talks, gave all her personal files on the negotiations to the National Archives, which promptly shredded them. "They were destroyed on purpose," Carney later reflected. "They wouldn't have been destroyed without someone's authority."[38] Gordon Ritchie, Canada's deputy chief trade negotiator, was "surprised" that the media focused on what he "regarded as a non-issue, the provisions on energy." His insider's account in *Wrestling with an Elephant* says nothing about who pressed the proportionality rule. He did "not see what all the fuss was about."[39]

Canada had little bargaining power in the FTA talks because before they began, Mulroney's government gave away all their aces in the 1985 "Western Accord" between Ottawa and the Western provinces. Canada gave up its ability to set a higher price on the oil and natural gas it exported, determine the price at which Canadians buy back their own energy from foreign transnationals, and conserve sufficient amounts for Canadians' future needs by restricting exports. But the Western Accord was a temporary policy that could have been overturned at the next election. To prevent this, proportionality and an energy chapter were placed in the FTA and then NAFTA.

After discarding its aces, Canada was the *demandeur* in the FTA talks. The treaty meant more for Mulroney's political agenda and prospects than for US president Ronald Reagan's government, so Canada's bargaining power was further weakened. Canada's aim was to get relief from US protectionism, while the primary US goal was to gain assured access to Canadian energy and end Canadian restrictions on US corporate ownership and control. The US won entirely on energy and partially on ownership, while Canada lost on gaining greater access to the US market, but got the face-saving device of ineffective trade dispute panels instead. Eighteen years after negotiating the FTA, Gordon Ritchie sounded despondent when writing about it in the *Globe and Mail*: "Canadian exporters, far from being guaranteed protection against the unfair application of US trade laws, are actually in a worse legal position than exporters from non-NAFTA countries."[40]

There was glee over proportionality in some Canadian quarters. "Critics say the problem with the [FTA] is that under its terms Canada can never impose another NEP on the country," stated Carney. "The critics are right," she continued. "That was our objective....If the Americans promise not to block our energy exports, we promise in turn not to turn off the tap on energy supplies shipped under contract."[41] Peter Lougheed was the FTA's most effective booster. He enthused that "the biggest plus of this [agreement] is that it could preclude a federal government from bringing in a National Energy Program ever again."[42]

Helmut Mach, Alberta's official trade representative during the FTA negotiations, confirmed that the separate energy chapter was added because of optics: "Alberta's objective was to maximize energy exports to the US. Alberta also wanted to allow oil and gas incentives for exploration. The NEP hung over the head of the major economic activity in Alberta at the whim of the Canadian federal government because of their power over exports. Alberta's objective was to never let a NEP happen again." But like Ritchie, Mach dismissed the proportionality clause's effectiveness in ensuring that there never be another Canadian NEP. The clause was technical, Mach argues. "Does anybody believe Canada would impose export restrictions in the next fifty years? So there was no thought that proportionality would have to be applied." Alberta played such a leading role in fashioning the FTA's energy chapter that some US negotiators and Canadian negotiators called it the "Alberta Chapter."

Once Mulroney's government had signed on to proportionality, they promoted it strongly to Mexico during the NAFTA talks. Carlos Salinas, Mexico's president, wrote: "It was extremely significant that Canada had accepted a [proportionality] clause ensuring the supply of oil. With that precedent, the Canadians argued that if Mexico did not agree to the same thing, in the current NAFTA, they would threaten to retract the clause from the bilateral agreement [FTA]. That would force the United States to put enormous pressure on us."[43]

1993 Election on NAFTA

The 1980s and 1990s were an era dominated by nostalgia, a yearning to return to the good old days of cheap plentiful oil, capitalism unfettered by governments and regulation, and the return of US power after the Vietnam War debacle. It was a brief historical lull of corporate rights romanticism and endless talk of a borderless world between the 1970s oil crises and the current debates around peak energy, climate change and border security. Corporate elites viewed Canada's problem as the opposite of energy sovereignty. Their issue was how to get assured entry into the US market for "excess" Western Canadian oil and natural gas during a temporary glut. This thinking and these interests placed the energy proportionality millstone around Canada's neck.

In the 1988 election, a slim majority of Canadians voted to defeat the FTA, but divided their votes between the Liberals (32 percent) and the NDP (20 percent). These voters got their chance again in 1993. It was the Liberal Party's first trip to the polls after their fierce opposition to the FTA in 1988. The Liberals were wary of unreservedly endorsing NAFTA, the successor deal to the FTA. NAFTA was signed and ready to start in January 1994, two months after Canada's 1993 federal election. There was still time for a new Canadian government to reopen or reject NAFTA. In their "Red Book" campaign platform, the Liberals promised that if elected they would "renegotiate both the FTA and NAFTA to obtain [among other things] the same energy protection as Mexico."[44] This was the Liberals' way of saying end proportionality.

The Liberals won their majority on the pledge not to sign NAFTA into force unless they got changes to energy, water and US protectionist actions.[45] So Prime Minister Jean Chrétien had to appear to do something. He reportedly approached US authorities behind closed doors but was summarily rebuffed. Democratic President Bill Clinton refused to reopen NAFTA, especially on energy. "We will not weaken or renegotiate any energy provisions of the FTA or the NAFTA. Specifically, we will not allow the Canadians to opt out of the proportionality clause or to limit its coverage solely to oil,'" Clinton wrote in a letter leaked at the time.[46]

Bill Richardson, a prominent Democrat and supporter of NAFTA, said that if Chrétien revived talks on the pact before Congress voted on it in mid-November, "that will hurt us."[47] Thomas Axworthy, former principal secretary to Pierre Trudeau, taught at Harvard University then. In his view, a public intervention by Chrétien, with the Canadian election mandate to renegotiate in hand, could well have scuppered NAFTA: "In 1993, when Bill Clinton was fighting to get NAFTA passed by Congress and facing very stiff opposition, had the newly elected Chrétien even uttered the word 'renegotiate,' NAFTA's opponents in the US would have used Canada to defeat the legislation."[48]

Rather than push back against US resistance to change NAFTA, Chrétien's government issued the following "interpretation": "In the event of shortages or in order to conserve Canada's exhaustible energy resources, the government will interpret and apply the NAFTA in a way which maximizes energy security for Canadians. The government interprets the NAFTA as not requiring any Canadian to export a given level or proportion of any energy resource to another NAFTA country. The government will keep Canada's long-term energy security under review and will take any measures that it deems necessary to the future energy security of Canadians, including the establishment, if necessary, of strategic reserves."[49]

These were the protections Canadians needed. But observers disagreed on whether the statement was a facing-saving device or not. Don McRae was one of the few experts who believed it was not. Currently a business and trade law professor at the University of Ottawa, McRae argued in 1993 that Canada could feasibly maintain that the declaration on energy security was legitimate, given that this position was made clear prior to NAFTA's proclamation. The declaration is "an indication of what the government intends to do if a dispute issue arises under the agreement," McRae said. "Whether the US immediately disagrees, lets it rest without any comment or whether it seems to act in accordance with that interpretation, these are all things that a tribunal or a court or a panel might take into account." He also noted that these types of statements carry greater weight if they remain uncontradicted. They have. The lack of American protest over Chrétien's declaration could be interpreted as tacit approval.[50]

Other experts doubted that Chrétien's statement would stand up in a NAFTA dispute resolution tribunal. US trade representative Mickey Kantor bluntly said, "None of these statements change the NAFTA in any way." Jean Charest, then one of the two Progressive Conservative MPs elected in the 1993 federal election said Chrétien "came back with zero, zilch, nothing.... It's a complete sell out."

Chrétien's "interpretation" may have brought moral force to Canada's declaration to "maximize energy security for Canadians." It could also prove
useful for a future government to dust off and use when the next oil shortage
strikes.[51] Why was Chrétien's government so weak on this file? Several reasons were advanced at the time. Washington opposed renegotiation, so the
Canadian government faced a choice of accepting NAFTA as is or scuttling it
entirely. The Liberals were not the strong economic nationalists that the John
Turner Liberals were in 1988. Chrétien's government was moving toward
the neoliberal, unregulated capitalism model that disastrously triggered the
Great Recession of 2008–09. As well, speculation was rife that if Canada got
an exemption on energy, the US Congress would demand an end to protection
of Canadian cultural industries, such as magazines, television, books and
films. Energy-producing provinces and Big Oil were also adamantly opposed
to altering or ending NAFTA. The main reason Chrétien was such a pushover
on renegotiating NAFTA was that in turfing the Progressive Conservatives
so soundly, voters inadvertently gave the Liberals too many seats. Majority
governments forget election promises more than minority governments. The
next election was four years off and the Liberals hoped that voters would
forget the issue by then. They did.

Obama Threatens to Pull Out

The issue of renegotiation came to life again during the 2008 US presidential
election. Barack Obama and Hillary Clinton both pledged to renegotiate NAFTA
to bring in tougher environmental and labour standards during the Ohio
primary for the Democratic Party's presidential nomination. To show she was
serious, Clinton asserted, "We will opt out [of NAFTA] unless we renegotiate
the core labour and environmental standards—not side agreements, but core
agreements." Later, she went further. If elected she would pull out of NAFTA
within six months of winning the White House.[52] Not to be outdone, Obama
raised the rhetorical bar: "We should use the hammer of a potential opt-out
as leverage to ensure we actually get labour and environmental standards
that are enforced."[53]

By threatening to use NAFTA's opt-out provision to bully Mexico and Canada into accepting the American way on altering NAFTA, Obama and Clinton
inadvertently reminded Canadians and Mexicans that their country could
leave NAFTA simply by giving six months' notice. What had been a fairly minor
American story from Ohio became much louder as it reverberated back
and forth across the border. The prospect of an Obama presidency sparked

a lot of "will he or won't he?" worry among the business-as-usual crowd in Canada. Would Obama really reopen NAFTA? How could the Canadian government help convince him not to? Canada used the energy card to try to bully Obama into withdrawing his opt-out threat. By putting energy on the table if Obama forced NAFTA renegotiation talks, Harper unintentionally gave Canadian critics an opportunity to reopen the proportionality debate.

David Emerson, then Canada's trade minister, warned that "if you open [NAFTA] for one or two issues, you cannot avoid reopening it across a range of issues." He added that "Americans' privileged access to Canada's massive oil and gas reserves could be disrupted if Washington cancels the NAFTA accord as Democratic presidential candidates threaten." Harper made a similar, if more subtle, threat in New Orleans in April 2008. But by telling Americans they have a sweet deal with Canada on energy, Canadian officials raised the question of whether Canadians got a sour deal.[54]

Once he wrestled the Democratic nomination away from Clinton, Obama began to listen more to US corporate power than to blue-collar voters in Ohio, whose demand for protection from cheap Mexican labour had pushed him into making the rash anti-NAFTA promises. Obama softened his renegotiation stance, saying his rhetoric had gotten overheated and he wouldn't unilaterally withdraw from NAFTA. Harper was overjoyed. "Once we reopen it," he observed, "it would be very hard to get that cat back in the bag."[55] He was right. Once the myth of NAFTA's inviolability is breached it can easily be breached again. But just six weeks later, Harper advocated just that. Irked by a growing "Buy America" campaign that saw many US cities and states refuse bids from foreign companies, including those from Canada, Harper campaigned to create "a new trade deal" with the US to put the awarding of local contracts under a continental free trade umbrella.[56]

To avoid appearing to contradict his previous opposition to reopening NAFTA, Harper called it a new deal rather than a renegotiation. By the time he became president, Obama was ready to comply fully with the petro-elites while still talking a good green game. Obama's message that the Sands are a necessary evil had become realigned with that of Ottawa, Alberta, Big Oil and the corporate media: even if it's dirty, the US must take it, as the alternative is to rely on Iran or Venezuela. This is faulty logic that conveniently ignores conservation: if the US reduces its per capita oil use to French or British levels, it can stop importing oil altogether and still use twice as much as the world average.

Playing the US Game

Cyndee Todgham Cherniak, a top international trade lawyer in Toronto, describes an ingenious way to reduce the proportion of energy exports Canada must make available to the US while staying in NAFTA. It would not require a tricky round of NAFTA renegotiations in which the US, if it agreed to let Canada out of the proportionality obligations, would surely demand that Canada give up something really major. Canada could play the same hardball game the US used against Canada on softwood lumber, Cherniak argues: create a bogus case for reducing Canada's energy exports and then delay a US proportionality challenge by dragging out the process for selecting the NAFTA panel to hear the case. Selecting a mutually agreed-upon chair can take a long time. If Canada drew out the process beyond three years and then lost, it could still win, since Canada's required energy exports are based on their share in the past three years. The clock keeps ticking. When proportionality is redetermined after Canada loses this hypothetical panel case, the energy share Canada must make available for export to the US would have dropped a lot. When Canada loses the case, Cherniak says, "Canada can say 'We made a mistake. Sorry, bad Canada.' But, we protected our people for three or four years. Canada should play the US game. Why shouldn't we do it?"[57]

It's intriguing, but it's very unlikely to happen because of the power imbalance between the US and Canada that lies at the heart of what's wrong with NAFTA. Cherniak did not suggest this, but to pull it off, Canada would need to elect an NDP or Liberal government, adopt Yankee-style confidence and fend off US tantrums. Washington takes US national energy security very seriously. It would also require a majority government's four-year mandate. If a minority government tried to pull off Cherniak's scenario, it would likely fall before the three-year period of declining Canadian oil exports to the US was up. The successor government could well cave in to US demands that Canada abide by its existing proportional exporting obligation.

Besides, if timid Canada suddenly got the moxie to force down virtually mandatory energy exporting to the US the way Cherniak outlines, it would also have the courage to demand that NAFTA renegotiation talks be held. Canada could insist on getting a Mexican-style exemption on energy proportionality as a non-negotiable demand. If the US and Mexico refused, Canada would have the grounds and Canadian public support to use the six-month exit rule to leave NAFTA. Exiting, then lowering exports and redirecting domestic oil and natural gas to serve Canadians first, would present the next Canadian government with a *fait accompli*. If you are going to be bold, you may as well go all the way.

Chapter 5

Alberta: Fossil-Fuel Belt or Green Powerhouse?

The Stone Age didn't end for lack of stone, and the oil age will end long before the world runs out of oil.

—Sheik Yamani, former oil minister, Saudi Arabia

Thirty years out, we won't be burning hydrocarbons the way we do today. Our enemies may not be at the door yet, but they are beginning to circle around Alberta.

—Clive Mather, former CEO of Shell Canada, 2011

When United States President Dwight Eisenhower appointed "Engine Charlie" Wilson as his Secretary of Defense, Wilson, who headed General Motors, was asked if he could make decisions adverse to General Motors' interests. He replied, "Yes, but I can't conceive of such a situation because for years I thought what was good for our country was good for General Motors, and vice versa." Sixty years later, the federal Conservative government held a similar blinkered view. Except the Sands, not autos, have run this country.

"What's good for the Sands is good for Alberta and Canada" has been the refrain. The petro-elite's perceptions are so distorted they can't conceive of a Canadian public interest separate from that of exporting as much Sands oil as possible. They won't countenance it either. When the Harper government brought in anti-terror legislation (Bill C-51) that identifies the undermining of the economic or financial stability of Canada as terrorism, the RCMP labelled the "anti-petroleum" movement a growing and violent threat to Canada's security because they "are opposed to society's reliance on fossil fuels."[1]

It is dangerous to place all one's eggs in a non-renewable resource basket. Premier Rachel Notley, like Peter Lougheed before her, recognizes this. But if Alberta continues to rely mainly on the Sands, it may well suffer a fate like the auto rust belt in Michigan and southwestern Ontario. In the 2030s, people will shake their heads about the folly of Alberta having madly excavated its way down into a "fossil fuel belt," while everyone else stopped buying or shipping dirty Sands oil and moved on to a low-carbon society. Alberta's economy will be left with little but the detritus of closed Sands projects and leaking tar pits. This is not Alberta's inevitable future. The Alberta Premier's

2011 Council for Economic Strategy acknowledged the danger of Alberta failing to diversify its economy. It warned that "the creation of an affordable, environmentally friendly alternative to oil would be a great thing for the world. It could be economically devastating for Alberta if, when it happens, we are still heavily dependent on oil exports."[2]

A Gap the Sands Can't Fill

The International Energy Agency says the world needs to add 4.5 to 5 million barrels a day of new oil *each year* to replace declines in old fields. Alberta's Sands have been touted as a "game changer," a major new source of oil that will enable the US to get off Middle Eastern oil and achieve energy security. Are the Sands that important? What would happen if we took them off the table?

In 2015, the Sands' output was 2 million barrels per day, about 2 percent of the global total. The Canadian Association of Petroleum Producers (CAPP) forecasts Sands output at 4.8 million barrels per day by 2030.[3] I doubt that the Sands will reach that level, because of opposition to importing "dirty oil," limits to pipeline take-away capacity, the possibility of hard international caps on carbon emissions, shortages of water and eventually natural gas, and the high cost of Sands production. But if we accept CAPP's forecast, 5 percent of expected world output is just enough to replace a year of world oil depletion. The Sands then cannot deliver enough oil to appreciably prolong the age of easy oil.

Political obstacles to the shipment of Sands oil are rising within Canada. Citizens' movements to stop pipelines to the Pacific and Atlantic coasts have mushroomed in the past few years and are starting to influence provincial governments. The strongest broadsides are likely to come from international refusals to buy Alberta bitumen. Uncertainty caused by the oil price crash of 2014 is slowing Sands' expansion. New Sands projects already partly built are proceeding, but up to $60 billion in new projects were put on hold after the price slump. In February 2015, the International Energy Agency cut its growth forecast for the Sands by 430,000 barrels per day, or 35 percent, saying they will not revive quickly even if oil prices rise.[4]

A fall in world oil production will spike oil prices and curtail globalization as "the death of distance" for trading goods. Low wages in Asia will no longer compensate for much higher transportation costs. It will not be feasible to import many goods from the other side of the world. Jobs that were exported in the past three decades will return to Canada. Inwardly directed development will become the new paradigm again. What was old will be new again. The question is whether and how quickly Alberta will get off the export mentality and embrace the new.

Can the Sands Be Greened?

The Sands serve no urgent purpose, as they produce mainly for export. The US doesn't need Sands oil. A *New York Times* editorial argued that the "Keystone XL pipeline is not only environmentally risky, it is unnecessary" because there is already sufficient pipeline capacity to double US imports from Canada.[5] The US wastes more oil than it imports. The US doesn't need Canadian or any other foreign oil. Sending Americans the dirtiest of Canadian oil further raises carbon emissions.

We've seen that Canada can't let Sands production grow as forecast and still reach Parliament's 2008 pledge to cut greenhouse gas emissions by 80 percent by 2050. Ramping up Sands output by three times while lowering overall emissions by 80 percent means the Sands would take up all of Canada's emissions room. Canadians would have to stop heating their homes and driving to work to reach the target. Sands extraction is shifting from mining toward deep in-situ methods, requiring more natural gas use and accompanying emissions. The business-as-usual scenario is for Sands output to rise as forecast, with Canadians continuing to use twice as much oil per person as Swedes and Britons.[6] If so, forget the "dirty oil" label; Canada would become the world's environmental rogue state.

The Government of Alberta website states that Canada releases only 2 percent of the world's GHGs, which is true but misleading. At 2.3 percent of the world's emissions, they're more than four times Canada's 0.5 percent of the world's people. With 12 percent of Canada's people, Alberta produces 30 percent of its emissions, four times the per-person level of the rest of Canada.[7] Only Saskatchewan's per-person emissions are as bad. The Sands can't be greened using current technology at reasonable cost. Their production emits much more GHGs than conventional oil.

The US Environmental Protection Agency estimates that Sands emissions are 82 percent higher on a wells-to-wheels basis. That's not counting destroying swaths of the boreal forest and wetlands that cover the Sands and sequester large amounts of carbon, nor heating bitumen to move it through pipelines. Emissions on an extra 900,000 barrels per day on the Keystone XL pipeline would be 27 million metric tons, "roughly equivalent to annual

CO_2 emissions of seven coal fired plants," according to the Environmental Protection Agency.[8] Other estimates peg the Sands' extra carbon emissions both lower and higher.[9]

Carbon capture storage was the comeback for Alberta Conservative governments. Let the Sands grow but capture much of their emissions. Have your cake and eat it too. The plan never made much sense and Big Oil wasn't interested because carbon capture storage is more costly than releasing carbon. The World Wildlife Fund estimates that carbon capture storage may capture 90 percent of the carbon emitted by new, coal-fired electricity generators, but only 3 to 15 percent from the Sands. Carbon from the Sands is too diffuse to capture effectively.[10] Alberta's carbon capture plan was a very expensive public relations reply to cries of "dirty tar sands oil." The Notley government sensibly plans to kill carbon capture and transfer the funds to public transit. The latter will reduce more emissions but, like carbon capture storage, will do nothing to reduce the Sands' growing emissions. A much better solution is not burning Sands carbon particles in the first place.[11]

Importing countries are becoming likely to refuse to take "dirty" Sands oil. The European Union has a fuel quality directive to reduce the life cycle of carbon emissions on vehicles. Although Europe imports virtually no Canadian oil, the EU has seesawed about slapping a dirty fuel label on Sands oil. The debate puts pressure on European oil giants Shell, BP and Statoil to divest their Sands holdings.[12] With its Low Carbon Fuel Standard, California may curtail Sands oil imports.[13] The provinces, including Alberta, should adopt similar standards on oil. Ottawa has no Plan B. It's wiser to embrace the coming low-carbon economy than to gamble on dirty oil exports continuing.

De-globalization Is Coming

Ottawa and the oil corporations don't see de-globalization as a possibility. Prime Minister Harper expects Alberta's Sands to turn Canada into a global energy powerhouse. For those of this mindset, a soaring world oil price would be a golden opportunity to realize the full potential of the high-cost Sands. Even if pricey oil causes global economic turmoil and reduced growth, they assume Alberta would have a long boom, fuelled by high oil demand. Canada possesses "the most attractive combination of circumstances for energy investment of any place in the world," Harper has said.[14] Only a massive dose of reality or political defeat will get Canada's petro-elites out of their dream world.

Alberta and Canada are currently caught between two powerful forces: the pressure to increase Sands output and rising counter-pressures to cut CO_2

Carbon Capture Storage (CCS)

70%

Alberta's Climate Change Strategy calls for a reduction of 200 million tonnes of carbon emissions by 2050 with 70% of cuts attributed to CCS, a commercially untested solution.

140 times

Alberta proposes to use CCS technology for a project that is 140 times as large as a North Sea facility that strips carbon from natural gas for commercial reasons.

90%

It's estimated that CCS may capture 90% of carbon emitted by new coal-fired electricity generators but only 3% to 15% from the Sands.

Carbon is likely to leak into the air and contaminate drinking water.

Separating CO_2 from flue gas for CCS involves large energy losses.

CCS guarantees increased capital and energy costs to build CO_2 pipelines, drill injection wells and monitor storage sites.

CCS could cost up to $14 billion a year, or a third of Alberta's annual revenues.

Big Oil is not enticed by the $2 billion Alberta offers corporations to adopt CCS.

CCS is expensive PR and gives Alberta an excuse to do nothing.

emissions. Deep conservation—using less carbon fuel—is the only realistic
way out of the dilemma. If they were truly "conservative," governments that
call themselves Conservative would embrace conservation and thrift. Even
some British Conservatives get it, or claim to. The party's Quality of Life
Policy Group noted that "material prosperity has not made us a contented
society," and "beyond a certain point material gain can become not a gift
but a burden." The group began its "Blueprint for a Green Economy" with a
1756 quote from founding conservative philosopher Edmund Burke, who
criticized the insatiable pursuit after more.[15]

Lloyd's of London, a risk assessor for three hundred years, advises countries and companies to prepare for disrupted energy supplies and costly oil.
A Lloyd's white paper on sustainable energy security states that "the more
efficient will have an important competitive advantage in times of high and
volatile energy prices."[16]

But miracle seekers keep looking for new technologies to reverse the
depletion game. Richard Heinberg explains how the imminent scarcity of
easy oil and natural gas once we burn through shale gas's temporary boom
is often misportrayed as "running out": "What we are really talking about are
the inevitable consequences of ... resource extractors [taking] the low-hanging
fruit first and [leaving] difficult, expensive, low-quality and environmentally
ruinous resources to be extracted later."[17] We have built a society, Heinberg
continues, "on the basis of *cheap* energy and materials.... As we move down
the layers of the resource pyramid, rising commodity prices and increasing
cleanup costs...will undercut both demand for resources and economic activity
in general." Labour costs will fall and raw materials rise in this century, the
opposite of twentieth-century trends. It will take time for people to realize
that "conventional economic growth is over. Done."

Globalization is built on cheap oil. Its end will mean relocalizing and renationalizing society. Distance will matter again and act as a huge tariff-like
barrier against distant products, as it has for most of history. What will the
economy look like when oil prices soar again and—after sharp gyrations and
countertrends—stay aloft? As Jeff Rubin puts it, "The world's cars and trucks
and ships and planes run on oil. That means the global economy runs on
oil, because the global economy is about moving things around the world.
And the reason the global economy has put all its eggs in one basket is that
there is no other basket....When the price of gasoline goes up, you drive less.
When the price of clothes or computers or anything else goes up, everybody
buys less. And when everybody spends less, you have a recession."[18]

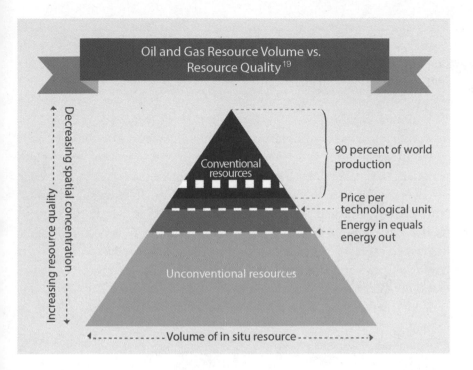

Rubin argues that many manufacturing sectors that recently deserted the Global North will return, creating an opportunity to produce things here that we started importing. It will especially be heavy, low-value products that costly transportation will relegate to local production. Wages in China are rising and will no longer offset long-distance transport costs. Rubin foresees the steel and furniture industries returning to Canada. Alberta won't likely get much of the steel industry, but should have a plan to get more of the conventional oil parts and equipment business. Canada and its provinces should strategize to attract the most viable industries that return from the low-wage Global South. High transport costs will also lead to relocalization and smaller firms, which by their nature have little capital. This shift will create many more jobs per dollar invested.

Dutch Disease

The Economist magazine coined the term *Dutch Disease* in 1977 to describe the effects of the Netherlands's natural gas boom on its manufacturing sector. The boom sucked in vast amounts of foreign capital, greatly inflated the Dutch guilder and priced Dutch manufacturers out of domestic and export markets. When does the phenomenon become a disease? Economist Paul

Krugman's answer is when manufacturing fails to return after the resource boom ends. Whatever the term, the concept has been long accepted by the political right, centre and left. It became ideologically charged in Canada in the spring of 2012, when Tom Mulcair raised it in a *Policy Options* article during the federal NDP leadership race.[20]

Mulcair's reasoning had to be discredited because it challenged Big Oil's exporting interests and the Harper government's intention to make Canada an energy superpower. What better way to do so than by trotting out Western regional grievance? Dutch Disease was catapulted from obscure academic discourse to Canadian public debate when Brad Wall, Saskatchewan's right-wing premier, attacked Mulcair for blaming the loss of manufacturing jobs on the Sands' role in boosting the Canadian dollar. To reduce the Sands' destructive effects on the environment and Canadian manufacturing, Mulcair advocated a cap-and-trade system to force Sands corporations to internalize environmental costs.[21] Wall said Mulcair's "make the polluter pay" call was code for a carbon tax that would redistribute wealth from Western Canada to the East.

To see if Canada suffers from Dutch Disease let's look at whether 1) the Sands boom substantially raised the loonie's value, 2) the inflated currency hurt manufacturing and other sectors, and 3) manufacturing's decline can be fully explained by other causes. No one argues that Canadian manufacturing, which shed over half a million jobs, or 20 percent of its workforce, in the decade ending in 2011,[22] was harmed only by Dutch Disease. But did the Sands boom and the high-flying loonie hurt manufacturing and other sectors? Will these sectors bounce back now that oil prices and the loonie have dropped?

First, did rapid Sands growth after 2002 boost the loonie's value? Mark Carney, former head of the Bank of Canada, acknowledged that strong crude oil prices forced the loonie's rise and that in turn hurt Canada's manufacturing exports; but he also declared that the loonie isn't a petro-currency. However, the Bank of Montreal's deputy chief economist Douglas Porter shows that the loonie traded in tandem with commodity prices 93 percent of the time in the decade after 2001, when the Sands were the main contributor to the raised value of Canada's commodity exports. The loonie traded at US $0.61 in 2002. It was a bad time to tour the world, but a good time to find work here. After 2002, prices for Canada's resource exports soared, boosting the loonie. Was it coincidence that manufacturing shed 354,400 jobs in the next five years, while Canada added over 1.4 million jobs overall?[23]

The 1990s, before the Sands boom, were good times for a diversified economy. For the first time in history, Canada "produced about as much manufactured output as we consumed"[24] and resource exports dropped to less than half (35–40 percent) of Canada's exports. It looked like Canada had finally broken out of its "staples" or resource exporting trap. "Made in Canada" was competitive enough then that exports of all kinds—manufactured goods and resources—reached almost half of Canada's gross domestic product (GDP). Meanwhile, foreign ownership and control of corporations in Canada fell to post-1945 lows.[25] Gains were partly realized from the Canadianization policies of the 1970s and early 1980s, before the effects of the Canada–US FTA and NAFTA fully took hold. Things were good for Canadian workers.

Second, world economic shifts around 2000 abruptly reversed Canada's diversification. China's rise led to an international commodities boom, and prices for Canadian resources soared. Blocked by corporate rights agreements like NAFTA from developing an inwardly directed economy, Canadians are once again "hewers of wood." Exports have fallen from almost half to less than a third of Canada's GDP and resources again make up nearly two-thirds of our total exports. Canada had caught Dutch Disease. Negatives outweighed positives. Manufacturing, forestry and agriculture shrank much more than resource exports grew. Even Alberta lost 20,000 manufacturing jobs and 18,600 agricultural jobs as the loonie rose between 2002 and 2007. The loonie hit US$1.10 by 2007, traded around par until 2013 and then fell to about $0.80. Why did the loonie suddenly rise after 2002? It was "less the resource boom than the financial boom associated with it," argues economist Andrew Jackson. Takeovers of oil, gas and mining companies boosted the loonie's value more than the export of those resources did.[26]

Third, some economists blame Canada's manufacturing job losses on "China Syndrome," a structural shift in the Global North from manufacturing to services as industry departed for China and elsewhere in the Global South.[27] China's cheap goods deeply penetrated Canada, hurting Canadian textiles, clothing, forestry and especially autos, a StatsCan study reports.[28] China's threat to Canadian manufacturing is a valid, partial explanation. But it is receding as China pays its workers more and lets the yuan rise in value. Canadian manufacturing was also hurt by the loonie's fast rise, independent of China Syndrome.

The Harper government vociferously denied that its promotion of the Sands as Canada's economic engine hurt manufacturing and other sectors. But well before Mulcair's article, Industry Canada commissioned economist

Michel Beine and his colleagues to study Dutch Disease. They calculated that the loonie's rise from 2002 to 2007 led to 33–39 percent of Canada's manufacturing losses of over 300,000 jobs then.[29] US economic weakness caused the remaining manufacturing job losses, they argued. So while constructing new Sands projects caused a jobs bonanza, it likely killed many more jobs than it created. Perhaps the Sands are more Canada's economic brakes than its engine.

Economists obsess about exports, touting *more* as always good. But are they? A country that exports more must also import more. *What* it exports and imports is more important than its export level. If it exports mainly raw resources and imports mainly manufactured goods, high tech products and services, it's exporting tonnes of jobs. Why? Because except for agriculture and forestry, resources are very capital-intensive and produce few direct jobs. Canada's entire oil and gas industry, extending beyond the Sands to conventional oil in Alberta, Saskatchewan and Newfoundland, and including natural gas across Canada, directly added only 16,500 jobs between 2001 and 2011.[30]

Canada's oil and gas sector employs half a worker for every million dollars of GDP, compared to manufacturing employing ten workers per million— twenty times as many jobs per unit of GDP. Construction creates eight jobs per million dollars of GDP. So constructing new Sands projects generates many jobs in Alberta. Statistics don't allow us to separate Sands construction jobs from general construction jobs in Alberta. But during the Sands boom from 2002 to 2012, total construction added 80,000 jobs in Alberta, a boost of 57 percent. The Sands undoubtedly created many of these gains. As long as new Sands projects are being built, there are many construction jobs. But when that phase ends and production begins, most jobs will disappear. When new Sands projects cease because of delays in new investments caused by the oil price crash, boycotts, blocking of new pipelines or enforcement of tough environmental regulations, the Sands' job boom will vanish.[31]

Putting all of Alberta's eggs in the Sands basket is a staples trap. It's also a "carbon trap," as Brendan Haley eloquently argues. Canada is the international greenhouse gases outlaw. Harper reneged on Canada's international, legally binding promise in 1997 at Kyoto to cut carbon and other emissions by 6 percent from its 1990 level by 2012.[32] In 2007, Ottawa changed that to reducing Canada's overall emissions by 17 percent from 2005 levels by 2020. Canada cannot possibly meet even that watered-down target, and allowing major Sands expansion even at a scaled-back level is the biggest reason. A 2014 Environment Canada report showed that oil and natural gas production in Canada surpasses transportation as Canada's biggest source of GHGs.[33]

The Dutch Disease debate is so contentious because manufacturing jobs and Sands jobs largely occur in different parts of Canada. Over 90 percent of manufacturing job losses between 2002 and 2007 happened in Central Canada, with over 50 percent in Ontario and almost 40 percent in Quebec. Alberta's manufacturing job losses were 6 percent of Canada's total.[34] Meanwhile, most Canadian jobs connected with the Sands are in Alberta. Oil job gains in Alberta and manufacturing job losses in Central Canada could set the stage for a regional battle royal over control of energy, environmental protection and economic rents. The latter are the gain in value of resources claimed by owners.

An Alternative Future for Alberta

Regional battles over energy and the environment loom. Who will win and who will lose when Canada shifts to a low-carbon future? If Albertans see a positive, job-creating plan that gradually weans their province off the Sands, they will react differently than if they see themselves as the big losers. Peak oil, general resource depletion, climate change disasters and energy security in a post-9/11 atmosphere cast new light on national unity and regional issues.

"Shut down the tar sands," says Greenpeace. While I agree with the sentiment, it would be a disaster if it happened right away. The Sands are too central to the economy and lives of Albertans. The challenge is to convince many Albertans and the Notley government that the present course is no longer viable, that phasing out the Sands is pro-Alberta and pro-Canada.

The Premier's Council, a group of corporate executives and former Conservative cabinet ministers, urged Alberta in 2011 to plan for a post-Sands economy. "We must plan for the eventuality that oil sands production will almost certainly be displaced at some point in the future by lower cost and/ or lower-emission alternatives," the group warned.[35] Despite the council's pedigree, the call fell on deaf ears. The stick-with-the-Sands crowd still runs Alberta's economy and may well tame its NDP government.

After flying over the Sands in 2006, Peter Lougheed, the man who started Alberta's Conservative dynasty (1971–2015), remarked: "I was just up there on a trip, just helicoptering around, and it is just a moonscape. It is wrong in my judgment, a major wrong, and I keep trying to see who the beneficiaries are. Not the people in Red Deer, because everything they have got is costing more. It is not the people of the province, because they are not getting the royalty return that they should be getting."[36] Lougheed also said it was time to "consider an increase in corporate and personal taxes" and orderly development—not more than two Sands projects at a time. And if the oil companies didn't like that, he stated, "we are the owner and we have the mandate to do that."[37] Lougheed advocated processing as much bitumen as possible in Alberta.

The Alberta NDP's "Green Energy Plan" calls for similar things. Mandating that "at least the value added/upgrading for all bitumen mined in Alberta be done in Alberta," the NDP also calls for a "Green Energy Fund" based on higher royalties to support renewable energy.[38] It's a strong environmental and pro-union variant of Lougheed's strategy. Why did Alberta's NDP adopt Lougheed's position? It's not that Lougheed shifted left after leaving the premier's office, but that the Conservative Party he led moved so far right. Equally, Canada's left-wing party, which used to advocate nationalizing the oil industry, something supported by half of Canadians as recently as 2005, has shifted rightward.

The Lougheed/NDP strategy of upgrading and refining resources in Alberta assumes that a narrowly based resource economy cannot afford to wait for markets or corporations to magically discover that Alberta's comparative advantage lies in diversifying the economy. Instead, they call on governments to lead in *creating* a comparative advantage beyond extracting raw resources. Lougheed put it this way: "You always have to keep in mind that we're the owner of the resource, the people. We should always be in a position where we could change the royalty rates.... [When I was premier, our government] would not give licences for oil sands development that were just in the mining side, but [would give licences] that required an upgrader. It's crucial to pace the boom to reduce inflationary pressures and get higher economic rents for the owners and the government."[39]

Economic Linkage

Lougheed describes the capitalist development model around a resource base that Mel Watkins, one of Canada's foremost political economists, articulated so well in 1963. The key to Watkins's "staple theory of economic growth"[40] is to develop three diversification prongs that are closely linked to the exported resources. The prongs are "forward, backward and final demand linkages". Rather than export resources only to import them back as finished goods, "linkage" means they are produced at home. In the past, Alberta and other Canadian resource-based regions implemented "forward linkage" strategies to force industry to upgrade resources to intermediate goods before exporting them. In oil it was around bitumen upgraders, refineries and petrochemicals. Jobs are created this way, but far fewer than if diverse consumer goods were made from oil, natural gas or wood.

Under the deregulation required for free trade, though, even limited policies of upgrading or forward linkage, were ended and the related sectors declined sharply. "Backward linkages" enable the resources to be extracted and can include road-building, pipe for pipelines, and machinery such as oil derricks and supersized trucks that remove the forest floor over the Sands. Alberta imports the vast majority of goods needed for this work. According to "Bitumen Cliff", a report by activist and scholar Tony Clarke and others, over $20 billion a year had been spent in Alberta on "machinery and equipment purchases, driven in large part by the enormous capital spending associated with bitumen developments.... Most of the heavy trucks used in the bitumen sands mining are manufactured by Caterpillar in the US—a company that just closed its only Canadian manufacturing facility....The end result is the emergence of a large and growing trade deficit in machinery....This trade deficit reached over $7 billion in 2011."[41]

Branching out from oil and natural gas could lead to the manufacture of everything made of plastics, as well as solvents, fibres, pesticides and coatings—a final demand linkage. Alberta never got far down that path, but had a successful petrochemical industry in the 1970s to 1990s based mainly around turning natural gas into intermediate goods like ethylene and propylene, mainly for export to the US, where they were converted into polymers (plastics), solvents, resins, fibres, detergents and ammonia. Most jobs built upon Alberta's oil and natural gas were exported. Alberta was stuck in a narrow, semi-industrial rut, dependent on external demand for its oil and natural gas.

Alberta lost much of its petrochemical industry. Its comparative advantage lay in low natural gas prices in Alberta based on limited pipeline takeaway capacity. That advantage vanished when Big Oil and natural gas interests beat out a pro-diversification alliance of the petrochemical industry and its workers to open a gas pipeline to Chicago in 2001.[42] Alberta's petrochemical industry contracted. Retreat from Lougheed's limited diversification strategy showed that an economy can slip back to a "pure staple economy" exporting its raw resources.

"Final demand linkage," the third diversification prong in Watkins's theory, is a market way for domestic industry to emerge around resource workers who are numerous enough to consume a broad array of locally produced goods and services. If the population grows large enough, it can sustain economic activities that have nothing to do with the original resource, thriving even when the resources decline. This is true diversification. Calgary and Edmonton are now cities with populations over a million—enough people to potentially sustain a diversity of sectors in ways that Saskatoon and Regina in neighbouring Saskatchewan cannot.

Mostly, though, Alberta's potential has not yet been realized. When the best money can be had in raw oil and natural gas, why bother with alternatives? The Lougheed/NDP diversification strategy depends too much on a passive market paradigm and runs the risk of resource depletion or boycotts against environmentally destructive extraction and upgrading. Even if successful, it will create workers and businesses whose economic interests lie in hindering the preservation of nature, reduced oil use, and the switch to wind, solar and deep geothermal to power up electric cars, trains and rapid public transit. I used to support this model, but now I think it's a dead end for Alberta and other provinces, because it bets that the age of easy oil and other carbon fuels will continue. It relies too much on resource extraction and ignores the enormous environmental damage of upgrading and refining bitumen.

A better way is deep diversification. Instead of relying mainly on exporting resources or anything else, governments become drivers and planners of a more "inwardly directed" economy. But we must ease ourselves off the old economy gradually, not drop it like a stone. True diversification depends on Alberta collecting high royalties and other economic rents and using them to fund the development of new sectors. Building bridges to the next economy will be pricey. The key is to wring more from the old economy as we transition off it by charging much higher royalties on Alberta's carbon fuels. Norway is the exemplar.

Two cheers for Norway

Amidst gloomy talk of an international "resource curse" and the autocratic, corrupting influence of oil, Norway stands out as the only bright light. Helge Ryggvik writes about the amazing kudos he and other Norwegian oil experts get when visiting other oil-producing countries.[43] Norway's oil policy is widely regarded, Ryggvik writes, "as the only successful example where a country, after discovering oil, has built a competent national oil industry, yet still has managed to maintain an egalitarian welfare state." With five million people, Norway took on Big Oil, asserted national sovereignty and got most economic rents, or "non-renewable depletion charges," to benefit Norwegians.

I give Norway two cheers, not three. Norway is impressive, but it has slid back from its best days, in the 1970s and 1980s, of standing up for Norway's people, the ultimate owners of its energy bounty. That's when Norway set up government-owned Statoil and created a thriving national oil-servicing industry from scratch. Norway's gigantic oil-pension fund resulted from those early victories. Norway's earlier sovereignty approach to water and forests[44] informed its oil stance in the 1970s. Labour's left-leaning government quickly took public ownership over Norway's offshore energy resources.[45] But lacking enough oil competence, technology and capital, Norway invited in foreign oil corporations to start things off. After a learning period, the government planned to Norwegianize the oil and natural gas industry under public ownership.

After US based Mobil Oil had operated the Statfjord field for fifteen years, Statoil, the new government-owned oil firm, took over in 1987.[46] In two decades Norway had become the world's third largest oil exporter. Like elsewhere in Europe, the generation of "1968 youth" radicalized Norway. But they had a more left-nationalist hue, like their counterparts in Canada who pushed Pierre Trudeau's government to set up Petro-Canada. In Norway's historic 1972 referendum, voters chose sovereignty over joining the European Economic Community. This decision coincided with a surge of economic self-assertion in the newly decolonized Global South, where one oil corporation after another was nationalized or newly set up.

In this context, Norway started Statoil and issued the ten commandments of oil that included envisioning domestic control to serve the national interest, fostering backward linkages to a thriving domestic oil service industry, gaining forward linkages by processing goods from oil, ending dependence on foreign oil supplies, and doing it all with ecological integrity. For Ryggvik the oil commandments were ways "to ensure 'national governance and control.'" Oil wealth was to be used to create environmentally friendly resource development and a "qualitatively better," more equal society. Local society was to be enhanced. Oil development was to be slow, avoiding uncontrolled growth typical of profit-driven Big Oil.[47]

But under neoliberalism's onslaught, starting in the late 1980s, Statoil was one-third privatized in 2001 and told to act like a for-profit oil corporation. Statoil now tries to capture other countries' economic rents, contradicting the commandment that natural resources benefit the whole community,[48] and pursues oil opportunities and quick profits abroad with the same zeal as ExxonMobil and Shell.[49] Nevertheless, Norway's early oil independence

Norway's Road to Energy Independence

 Laws are passed to protect resources from foreign ownership and excessive corporate profit.

Foreign control of resources, chemicals and mining reaches 40 to 80 percent.

 The National Insurance Scheme Fund (later The Government Pension Fund) is set up.

Citizens vote against joining the European Union and Statoil is established to nationalize the oil industry.

 Norway takes public ownership over offshore energy resources but allows US-based Mobil Oil to operate the Statfjord field while training Norwegians to take over.

Statoil takes control of Norway's oil fields.

 Neoliberalism movement grows and Norway becomes the world's third largest oil exporter.

Government-owned Petoro is established to manage all of Norway's petroleum licenses. Statoil is one-third privatized and now captures other countries' economic rents.

quest shaped its current international leadership in capturing and saving economic rents.

Norway's very generous welfare state is funded substantially by oil revenues collected and saved in its "Pension Fund." Several establishment voices advise Canada to follow Norway in saving economic rents, but not the rest of Norway's model. A C.D. Howe report by Leslie Shiell and Colin Busby calls for Alberta to save even more than Norway. *Our Fair Share*, a report commissioned by Alberta's Conservative government, counsels copying Norway in maximizing royalties. Norway captures most of the "economic rents" by combining government ownership with imposing very high economic rent charges on oil transnational corporations. Norway adds a 50 percent special tax for petroleum companies to a general 28 percent corporate tax base, to capture 78 percent of net petroleum profits from the private sector. While this rate looks good, Norway abhors giving away the other 22 percent of unearned profits. To counter this, Norway owns two-thirds of Statoil, and all of Petoro, which manages oil and gas licences. Norway collects *all* the gross rents from the forty percent of the sector that is publicly owned.[50]

Norway's example is laudable in many ways, but it is locked into supporting carbon-fuelled growth. Its oil fund removed fears that moving too quickly in carbon-fuel extraction would overheat the domestic economy, so Norway had a shorter, sharper oil boom than its go-slow commandments of acting environmentally responsibly originally envisioned.[51] Instead of sustaining lower output for much longer, oil production started down the slope to terminal decline in 2002.[52] Norway's huge oil fund also makes it a "rentier state." Classical economists from Adam Smith to Marx condemned rentiers on the grounds that no one should enrich themselves without contributing work. By investing in for-profit entities abroad, Norway contradicts its ethos of popular national sovereignty. Instead, Norway is a giant absentee owner without connection to the people and land in which much of its money holds power.

Rock-Bottom Economic Rents in Alberta

Many think of royalties as taxes. Any government fee must be a tax. Wrong. Private woodlot owners and musicians collect royalties. No one calls them taxes. When governments collect royalties they aren't taxes either. Royalties are one way to capture economic rents. Leases, ecological charges and corporate taxes are other ways. Government ownership of resource companies is the only way to collect all the rents.

"The oil sands are owned by the people," Peter Lougheed insisted. "They're not owned by the oil companies." He urged Albertans to "think like an owner" and levy their fair share of royalties. Governments in Canada own most of the subsoil resources on Crown land. In Alberta, the province owns 81 percent of the subsurface mineral rights, while the federal government owns 11 percent (including national parks and First Nations reserves), and 8 percent is owned by individuals or companies.[53] At 17 percent, private ownership is a little higher in Saskatchewan. The federal government has formal ownership in the territories and offshore but shares jurisdiction with territorial governments and ocean-side provinces through agreements.[54]

Saving rents matters too. *Globe and Mail* reporter Eric Reguly contrasts Norway's long view with Alberta acting greedily and deciding "that a drunken, blow-out dance party today was better than a string of candle-lit dinner parties down the road."[55] Most other oil-producing jurisdictions squander oil rents, too. Norway invests all economic rents to avoid "Dutch Disease." Otherwise Norway's kroner would spike and hurt domestic manufacturing and other sectors. Norway collects much more economic rent than Alberta, although its oil output is lower and its natural gas output and production costs are similar.

Norway has advantages, though. As a maritime country, its oil gets a higher price and its natural gas a much higher price than landlocked Alberta's. Higher prices give the state much more room to capture economic rents on the same energy output. Even so, Alberta has had huge rent potential but wasted it. Parkland Institute's Regan Boychuk showed that Alberta gave away $121 billion in "excess" pre-tax profits between 1999 and 2008.[56] Excess profits are unearned profits derived from the value of public lands—what's left over after the costs of exploration, development and operations and a normal rate of profit (10 percent) are subtracted. Instead of Albertans getting the economic rents, as the owners of the resources, the excess profits were handed over to Big Oil in an act of "misplaced generosity," according to Boychuk.

Norway gets much higher rents than Alberta, in large part because of higher taxes and especially its public ownership stake. Norway wiped out its national debt and created an elephant-sized oil pension fund worth C$1.1 trillion and growing—making Alberta's older Heritage Fund look like a mouse at $17.2 billion.[57] Alberta can't catch Norway. But if it starts collecting at Norway's level, Alberta could have substantial funds to finance the transition to a low-carbon economy.

Alberta's previous Conservative governments claimed that its economic rents, among the world's lowest, attract investment that would otherwise

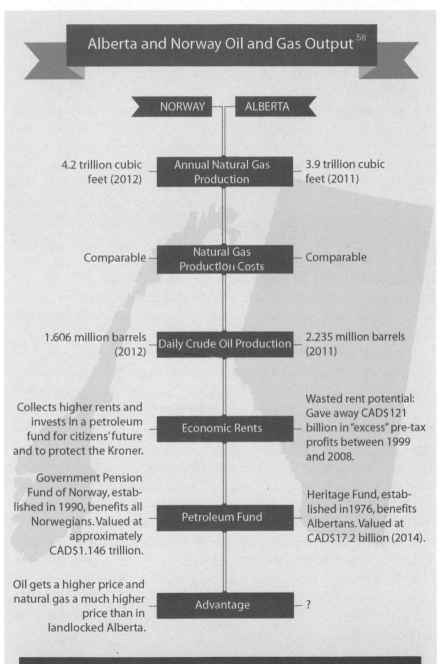

Alberta and Norway Oil and Gas Output[58]

NORWAY   ALBERTA

	NORWAY	ALBERTA
Annual Natural Gas Production	4.2 trillion cubic feet (2012)	3.9 trillion cubic feet (2011)
Natural Gas Production Costs	Comparable	Comparable
Daily Crude Oil Production	1.606 million barrels (2012)	2.235 million barrels (2011)
Economic Rents	Collects higher rents and invests in a petroleum fund for citizens' future and to protect the Kroner.	Wasted rent potential: Gave away CAD$121 billion in "excess" pre-tax profits between 1999 and 2008.
Petroleum Fund	Government Pension Fund of Norway, established in 1990, benefits all Norwegians. Valued at approximately CAD$1.146 trillion.	Heritage Fund, established in1976, benefits Albertans. Valued at CAD$17.2 billion (2014).
Advantage	Oil gets a higher price and natural gas a much higher price than in landlocked Alberta.	?

Conclusion:
If Alberta begins collecting rents at Norway's level, Alberta could have substantial funds to finance the transition to a low-carbon economy.

not come. This is false. Alberta and Canada have immense advantages over
many oil sites in the Global South: political stability, first-class infrastructure
and skilled workers. Jeff Rubin states that Alberta has 50 to 70 percent of
the world's oil still open to private investment,[59] which should be a great
bargaining chip.

Kuwait pioneered oil funds in 1953 to provide for future generations and
reduce reliance on a single non-renewable resource. Its fund is now over
$200 billion. Alberta was an early adopter, having started its Heritage Fund
in 1976, nineteen years before Norway.[60] Peter Lougheed's Conservative government set up the fund to spur economic diversity. Alberta put 30 percent
of its resource revenues into the fund, which reached its peak in 1987. At
$12.7 billion then, it would be over $25 billion in today's dollars, 50 percent
more than today.[61] In 1988, Alberta stopped adding resource royalty revenues to the fund and put all the fund's income into general revenues. While
much lower than they should be and wildly varying from year to year, direct
and indirect oil and natural gas revenues still fund one-fifth to one-third of
Alberta's government spending. Investment income from the Heritage Fund
adds another six percent.

Instead of relying on economic rents, Alberta should boost royalties and
put all of them into the fund. Like other provinces, it should pay for all
government services through adequate taxes. The Alberta NDP established
a Resource Owners' Rights Commission to review Alberta's royalty rates. If
the commission recommends raising royalties and the Notley government
concurs, the NDP has promised it will put 100 percent of the increased royalties into the Heritage Fund. That implies the province will leave revenues
collected at the existing royalty rates as general revenue.

The Heritage Fund lapsed because Alberta embraced the neoliberal idea
that the market, meaning large corporations, not governments, should decide
what Alberta's comparative advantage is. Alberta's laissez-faire approach is to
remain the lowest tax jurisdiction in Canada in the hope that it will attract
footloose industries. But this has never happened. Alberta is more than ever
an oil and natural gas province. Corporations from around the world invest
in Alberta because that's where the petro-resources are, not because Alberta
has no provincial sales tax.

False Spring

In 2007 it looked like Alberta might finally stop its giveaways to giant oil
transnational corporations. Ed Stelmach, Alberta's Conservative premier,

set up a panel to review the province's oil and gas royalties.[62] The panel's *Fair Share* report showed Alberta's take was far below that of Norway and Venezuela, and well below that of Texas, Wyoming, Colorado and California, jurisdictions known as Big Oil–friendly. Two-thirds of Albertans backed the report's call for higher royalties.[63]

Bill Hunter, former president of Al-Pac, a giant forest corporation in northern Alberta, chaired the royalty review panel. Hunter counselled Alberta to collect all the rents from its non-renewable energy resources. "As Albertans, we own 100 percent of the resource, and we should expect nothing less than 100 percent of the rent. It's up to industry to convince us that we should take a decrease."[64]

The panel's trenchant analysis contrasts with its timid recommendations, which would have left Alberta's royalty rates lower than those of the US states mentioned above.[65] But even so, powerful oil transnationals in Calgary saw red and counterattacked by trying to scare the public. You will kill Alberta's golden goose, we will leave, and Albertans will lose, was their message. After a year of record profits, a corporate CEO told the panel, "It's a myth out there that this is a hugely profitable business." But he confided to investors at the time that his company's Sands project will produce a "wall of cash flow sustainable for decades."[66] Meanwhile, Roland Priddle, former head of the National Energy Board, was in Texas to pitch Alberta as a great place to invest. "Where else can you purchase in-place oil [well, bitumen] for one cent a barrel?"

Stelmach blinked and timidly raised royalties only a little. Critics cried sellout. But amazingly, Big Oil aggressively attacked Alberta's modest royalty "hike" and moved rigs out of Alberta to scare people. While they peddled their hard-luck story, oil transnationals' take of economic rents from Alberta rose to new heights. Boychuk's report shows that oil and gas corporations' share of "excess profits" in Alberta rose from 35 percent in 2006 to 66 percent in 2008.[67] Despite this, Calgary's petro-elites were fighting mad. Many helped bankroll the Wildrose Alliance party, which promised rock-bottom royalty rates again. Stelmach quickly reversed course, met with the petro-corporations and excluded the public. To applause from the oil transnationals, Alberta dropped royalties. Fifty-eight percent of Albertans were opposed to this reversal,[68] a harbinger of the NDP breakthrough in 2015.

Expect a repeat scare performance by Calgary's oil patch in response to Notley's royalty review. Big Oil will argue that conditions are not right to raise royalties. Capital strike or capital flight may see oil rigs pull out and exploration decline. It could be a great show. Will Notley's government blink like Stelmach's did, or will it tough it out? As long as it shuns the economic nationalist and radical egalitarian underpinnings that made Norway's model such a success, Alberta and Canada will not collect and save rents at Norway's level. But only by doing so can Alberta readily fund its transition to a truly green economy.

Alberta's Green Conversion

Making Alberta a less energy focused economy can create many jobs. Out of fifty-six sectors, oil and natural gas extraction and mining are dead last in the creation of jobs for every million dollars invested. If much higher royalties are collected and put into health care and education, five to seven times as many jobs are created as would be if the same amount were put into oil and natural gas output. A green-energy economy is coming. Globally, it's already worth more than $4 trillion. Little Denmark is a leader in wind energy, which generates over 10 percent of its exports and 39 percent of its electrical power and has created 28,000 jobs. Germany's renewables sector employs over a quarter of a million workers. Good things can come to pioneers. Danish companies have installed over 90 percent of the world's offshore wind turbines. Alberta could still become a leader in specialized green energy technology and services, like ultra-deep geothermal power. But if it waits too long, Alberta will import most of its green infrastructure and expertise from abroad. That means exporting jobs and losing out on the transition off the precariously narrow Sands.[69]

What could Alberta's transition to a post-Sands economy look like? Independent public policy consultant David Thompson's report, *Green Jobs: It's Time to Build Alberta's Future*, provides a good preview.[70] He focuses on quickly reducing energy waste. Alberta's coal-based electricity generation emits almost five times as much greenhouse gas as Canada's average power generation, and releases pollutants that cause smog, acid rain, asthma, respiratory and cardiac problems, heart attacks and cancer, he writes.[71] Coal-fired power helps make Alberta an environmental pariah, by adding huge carbon emissions to the enormous levels caused by the Sands. Coal-fired power must quickly be replaced. Sands construction workers could be better employed retrofitting buildings and constructing new light rail transit, bus lanes with their own rights of way, and a high-speed train between Calgary and Edmonton. These changes would facilitate active transportation—transit, walking and cycling—and make neighbourhoods more vital and less energy intensive.

Many oil and gas workers will be laid off as we switch to a green economy. Thompson advises Alberta to fund retraining and financially support workers being retrained, like the American Recovery and Reinvestment Act did after the 2008–09 Great Recession, putting over a billion dollars into green jobs training. The Great Recession hit Alberta harder than it did other provinces. Full-time jobs fell by over 100,000 from August to December 2008 and were slow to return. Alberta nose-dived again after oil prices crashed in late 2014. Green jobs would be a very attractive alternative for Sands construction workers, because they are spread more evenly across rural and urban areas, enabling workers to live in their communities with their families rather than commuting to remote Fort McMurray, Thompson writes.[72]

Green jobs will come mainly in energy efficiency, renewables, transit and more labour-intensive, less toxic farming.[73] The quickest, biggest bang lies in public investment in energy efficiency and retrofitting buildings, which should be scheduled during the resource bust periods that always follow Alberta's booms, like in 2015. It's cheaper during these times and would give useful work to construction workers laid off from oil and gas projects. Public spending will kick-start most green jobs at first, but a mix of public spending and government policies that induce private spending will keep them going.[74]

Can the provincial government afford to pay for the greening of Alberta? Yes. Even after its great royalty giveaways and a major deficit in 2015 caused by falling oil and natural gas prices, Alberta is the only province with positive net financial assets: $8.8 billion. It also has large capital assets—$45.3 billion, which includes the Heritage Fund—so its total net assets are $54.1 billion.[75] Despite low projected energy prices, Alberta's assets are not expected to fall over the next five years. Alberta should not fall behind.

Framing it as going green to gain national energy independence, the US is reducing energy waste and curbing carbon emissions. US states and municipalities used nearly $8 billion in federal funds to upgrade energy efficiency. Washington aims to weatherize a million homes a year and improve energy efficiency in commercial buildings.[76] Japan and Korea also adopted green stimulus packages, each proposing to create a million jobs. Demonstrating that not all greening requires government spending, Germany boldly moved ahead with feed-in tariffs for electrical power. The grid pays more for energy from renewable resources and rolls those costs into consumers' power bills. To prevent those bills from rising too much, a massive insulation retrofit would lower energy usage. Following this example could mean lower natural gas usage, and perhaps lower costs for Albertans, to offset rising power rates.

Alberta lacks good hydro sites, but has excellent wind sites just east of the Rockies and good potential for solar. Thomas Homer-Dixon promotes ultra-deep geothermal power for Alberta: "We drill holes eight to ten kilometres into Earth's crust, pump down water, then bring it back to the surface—super-heated—to drive electrical turbines." Homer-Dixon also supports underground coal gasification, which, in contrast to burning coal on the surface, leaves it in the ground.[77] Alberta need not be self-sufficient in electrical power generation, though. Alberta sells oil to British Columbia and Manitoba. Why should those provinces not sell clean hydro power to Alberta instead of to the US?[78] That way, Albertans' power bills need not rise to get green power.

David Thompson contends that green jobs are good local jobs with a living wage and are "stable and less susceptible to volatile global commodity prices." Most are in fields in which people already work: construction, manufacturing, engineering and finance. Workers will get the added satisfaction of making a difference. A side benefit would be that more Alberta workers would likely welcome rather than block moves away from a carbon economy. Thompson projects tens of thousands of longer-term jobs in building more light rapid transit lines in Edmonton and Calgary that will achieve denser, truly transit-oriented cities, reduce commuting times and end urban creep onto good farmland. Two-thirds of Albertans support provincial funding of a high-speed rail link between Calgary and Edmonton. Once built, Alberta would join Europe, China and the US in replacing many medium-distance flights with high-speed rail. A rail trip from London to Madrid, for example, causes one-sixth the carbon emissions of a flight.[79]

How can Alberta afford so much new rail and rapid transit? Alberta spent $10 billion on highways and city roads between 2008 and 2011 to fund excessive auto use. Instead, Alberta could spend sufficiently on road maintenance and transfer most existing funds to new light rail transit and trains.[80] Sands construction workers could be redirected to these projects. About ten times as many workers build Sands projects as operate them. Most of the green economy work would be for a one-time switchover, though. You retrofit a building and put in a high-speed link between Alberta's main cities once. What kinds of jobs could sustain Albertans in a low-carbon economy in the long run?

Thompson proposes an Alberta Renewable Energy Corporation, a provincial Crown agency "to accelerate the development of renewable energy manufacturing capacity in Alberta.... The Crown corporation would immediately purchase inputs and begin to build the renewable manufacturing infrastructure."[81] It would operate on a commercial basis. This may seem like a radical idea, outside of Alberta's private enterprise tradition, but it's not. Alberta Treasury Branches, founded in 1938, is a government-owned bank that operates 164 branches and 133 agencies in 244 communities throughout Alberta. It's a very Albertan institution. The Alberta Energy Corporation was set up by Peter Lougheed in 1973 as a public-private vehicle for Alberta to directly participate in the Sands, though it was later privatized.[82] Using a Crown company to spark a green energy industry in Alberta is a good bold idea, but it needs fleshing out. Overall, the *Green Jobs* report relies too heavily on construction. Like construction in the Sands, a green retrofit will only temporarily employ Albertans.

After the Sands

It's crucial to avoid an abrupt shift to a low-carbon society. We must transition off the Sands and develop a vision and plan for what's next. The first step is to cap and then phase out the Sands over fifteen years, starting with the oldest projects. Meanwhile, new industries and jobs must be created around a green economy that builds on Alberta's highly educated and skilled workforce. Alberta's economy can be diversified. After the oil price crash of the 1980s, Texas rebuilt itself as a leading centre in aerospace, military equipment and computer technologies. After the steel industry collapsed, Pittsburgh transitioned to a hub for health care, robotics, banking and education. Alberta can build from its promising start in biotechnology, financial services, telecommunications, medical research and development, and environmental technologies.

Once moving off the Sands is underway, new economic sectors, workers and voters will emerge whose interests are best served by continuing the transition. The momentum could be sustained. An Alberta no longer beholden to Big Oil can benefit all Canadians immensely. Instead of blocking the greening of Canada, Notley's Alberta government could lead the way. "I can imagine an Alberta without oil, but not without water," Peter Lougheed observed.[83] There are far better alternatives for Alberta than to be stuck with its head in the Sands.

Chapter 6
Resource Nationalism Everywhere but Canada

I wasn't prepared to just give it away. Why should Canadians be second-class citizens and take less than others? The asset is in the ground. It isn't going anywhere.

—Danny Williams, Progressive Conservative premier of Newfoundland, 2006

People have to remember that the oil sands are owned by the people. They're not owned by the oil companies.

—Peter Lougheed, former Progressive Conservative premier of Alberta, 2008

The resurgence of resource nationalism is an overlooked story. Canada caught the 1970s wave, but is missing out on the current one. "Petro-Canada," "the National Energy Program" and "Canada First" were icons of the earlier wave. Now Canadian energy is portrayed as a continental resource. Why is control by public national entities winning out over foreign-owned oil transnationals in so many other countries, but not in Canada? What are the consequences of this for Canadians' energy security and conservation? Would Canada gain or lose by embracing twenty-first-century energy and environmental nationalism and a new form of ownership—*public interest* ownership?

When you bury a national icon like Petro-Canada, you must ritualistically wave the flag. "This will obviously be a company that is very focused in Canada, very focused in oil sands," stated Rick George, Suncor's CFO, when his company bought Petro-Canada in 2009.[1] George, an affable Coloradan, was right about the Sands focus, but is Canada at its centre? Why are Petro-Canada's remnants now helping burn through Canada's dwindling supplies of natural gas to heat dirty Sands oil for export? Was that why we used to fill our tanks with Petro-Canada's gas—to, as the ads told us, "pump [our] money back into Canada"?

While many countries are renationalizing the petroleum industry to ensure energy security, social benefits for citizens, environmental protection and national independence, foreign corporations dig up Canada's resources and ship them out raw. Canada's colonial "hewers of wood" pattern continues. Ottawa and the producing provinces get very low royalties and other economic rents from the permanent depletion of their non-renewable energy resources. But Canada is not immune to the renationalizing trend. There is much government-ownership of Canada's oil resources, particularly in the Sands. Most is recent. The problem is that none of it is Canadian.[2] Ironically, a Leger poll showed that half of Canadians favour "nationalizing" oil resources and companies in Canada.[3] Instead of responding to Canadians' wishes, the federal and Alberta governments allowed national oil companies from Norway, China, South Korea, Thailand and Abu Dhabi to own chunks of the Sands. Ottawa stopped the trend in 2012, but failed to replace them with Canadian national oil companies.

Canada is one of the few countries to shun public ownership. It wasn't always this way. After years of oil policy largely determined by big foreign oil, Canada boldly set up Petro-Canada as a Crown corporation in 1976 during an international wave of oil nationalism. Petro-Canada partly copied the example of BNOC (Britoil). France, Italy and Germany also had national oil companies. Petro-Canada bought several foreign oil transnationals to partly Canadian-ize the petroleum industry. Strong majorities in every province, including Alberta, supported the goal.[4] Canada cut exports of oil to the United States to 14 percent of their earlier level and redirected it to Eastern Canadians. Then as now, the US depended on insecure oil imports. The governments of Alberta and Ontario invested in the Sands.[5] In 1975, Peter Lougheed's government set up the Alberta Energy Corporation, partly owned by the province, to enable average Albertans to invest in their natural resources.

It's a shame Canada shuns the new resource nationalism. It's needed to guarantee oil for Eastern Canadians, combat climate change, protect local environments and conserve resources. Only public interest control ensures that governments collect all the economic rents from Canada's resource wealth on behalf of the owners—citizens, including indigenous peoples. Governments need the rents to fund the transition off carbon.

The New Seven Sisters

Once upon a time, the "Seven Sisters" of Anglo-American oil carved up the world in gentlemen's agreements and helped topple regimes hostile to their interests. In 1953, the US State Department suggested forming a "consortium" of the seven sisters of Anglo-American oil corporations to run Iranian oil after the CIA and Britain's MI6 helped overthrow Iran's elected government, which had nationalized Anglo-Persian, now BP. But oil nationalizations in the 1970s and mergers in the 1990s shrank the Seven Sisters to four—ExxonMobil, BP, Chevron and Shell. Although their power revived somewhat in the 1980s and

1990s with neoliberal privatization, Russia and Venezuela sparked a renationalization trend between 2000 and 2003. By 2013, national oil companies
controlled over 90 percent of the world's oil reserves.[6]

Six of the ten largest oil corporations in the world are now national oil
companies, with ExxonMobil only the fourth largest by production, Shell
seventh, BP ninth and Chevron tenth. The petro-giants with the largest oil
reserves are all national oil companies, each holding over 100 billion barrels.
They are Saudi Aramco, Gazprom (Russia), NIOC (Iran) and Kuwait Petroleum.
They make the old "Seven Sisters" look like Lilliputians. ExxonMobil has the
largest reserves of privately owned corporations, with 25 billion barrels—
compared with Saudi Aramco's 307 billion.[7]

The old supermajors now account for only a quarter of the world's exploration and production spending. And they are confined to costly, risky regions
like the deep ocean, the Sands and the Arctic. Costs are very high, but they've
few places left to go.[8] Alberta's Sands are one of the few places left for private
investable oil. This contradicts Big Oil's scare stories that oil transnationals
will leave if regulations and royalties are too onerous.

Economists versus Resource Nationalists

Why are privately owned oil corporations declining? Petro-elites and many
economists blame resource nationalisms, not Northern countries' colossal energy waste, for the coming world oil shortages. National oil companies are not
investing enough in new finds, they say. Outrageously, they charge, national
oil companies are leaving oil in the soil. Equally shocking to them is that many
national oil companies share their countries' resource wealth with their people.

Ray Walser, at the conservative Heritage Foundation in Washington, sees
a threat to US energy security in the "systemic assaults on the efficiency of
state-run energy companies to fund everything from social welfare programs,
government operating budgets, arms purchases, and aid to subsidize oil for
foreign friends." He's referring to Venezuela, leader of the "carnivorous left"
and practitioner of statist socialism, intensified resource nationalism and
anti-Americanism.[9]

Most economists think government ownership is a political risk and a
cost to foreign investors rather than a benefit to communities where the
resource is found.[10] The focus shows whose interests most economists deem
important. They invoke the magic of the marketplace. Left to work without
state interference, the market will, they argue, find new oil reserves, curtail
overuse and prompt substitutes. Some of this is true. Much is myth.

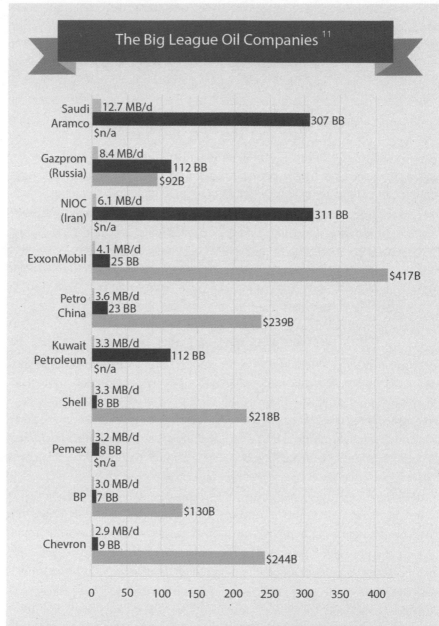

The Big League Oil Companies [11]

Company			
Saudi Aramco	12.7 MB/d	307 BB	$n/a
Gazprom (Russia)	8.4 MB/d	112 BB	$92B
NIOC (Iran)	6.1 MB/d	311 BB	$n/a
ExxonMobil	4.1 MB/d	25 BB	$417B
Petro China	3.6 MB/d	23 BB	$239B
Kuwait Petroleum	3.3 MB/d	112 BB	$n/a
Shell	3.3 MB/d	8 BB	$218B
Pemex	3.2 MB/d	8 BB	$n/a
BP	3.0 MB/d	7 BB	$130B
Chevron	2.9 MB/d	9 BB	$244B

0   50   100   150   200   250   300   350   400

Production, 2012 or latest, MB/d (million barrels per day)
Reserves, 2010 or latest, BB (billion barrels)
Market Value, 2013 (July 31), $B

Oil has rarely operated in free markets. After disastrous competition in Pennsylvania's oil fields in the 1860s, John D. Rockefeller's Standard Oil (a forerunner of ExxonMobil) brought the fields and much of the world into its monopoly. Oil became a strategic good fought over and guarded by militaries. It's been more monopoly than not since, dominated by the old Seven Sisters, then by OPEC and the national oil companies. When oil is an oligopoly or monopoly, who runs it and who benefits? Should profits and economic rents go to rich owners, be reinvested in finding new supplies, or support public services in oil-producing countries? It depends on who you think the oil belongs to. Is it the military and rich consumers in the Global North, or the countries and people who claim it as their patrimony? Can we leave environmental protection and people's access to dwindling carbon energy to unrestricted markets?

Resource nationalisms revived in reaction to the dominance of the neo-liberal Washington Consensus and US power. Proponents of freewheeling capitalism decry resource nationalisms. The more they spread, the more vociferously they are denounced. Robert Hirsch, author of a path-breaking report on peak oil, claims that "the greatest above-ground risk to future world oil production is almost certainly resource nationalism."[12] Before becoming US vice-president in 2001, Dick Cheney noted that governments and national oil companies control about 90 percent of the assets. "Oil remains fundamentally a government business."[13]

Economists assert that resource nationalisms move in cycles. As the international price rises, citizens in oil-producing countries feel they are being "ripped off" by foreign transnational corporations. Rising oil prices and resource nationalisms fuel oil-exporting states to get better terms with transnationals. When oil prices crash, the reverse happens. Producing countries are in a weak bargaining position. Economic rent collections fall. Resource nationalisms decline. To maintain exports and jobs, governments entice transnationals to stay or return by offering better terms. Transnationals then reoccupy the driver's seat. Not everyone accepts the oil-price thesis of resource nationalisms, because it can't explain the many instances where resource nationalisms triumph when oil prices are low. Nordine Ait-Laoussine, president of Swiss energy consulting firm Nalcosa, sees other root causes, such as the failure to convert revenues into modern social services, employment and higher living standards in oil-exporting countries.[14]

According to Paul Stevens, a petroleum economist and consultant, resource nationalisms have two components: "limiting the operations of International

Oil companies and asserting greater national control over natural resource development."[15] Motivation to achieve both goals remains strong regardless of oil prices. Stevens sees resource nationalisms as driven by concern that transnationals are taking too large a share and a perception that the resource will be needed domestically, that export customers are unworthy and that ordinary people receive little benefit from "their" oil.[16]

Which approach is right? Many observers note the rise of oil nationalism after the international oil price quadrupled in 1973, and its recent revival when oil prices rose sharply after 2001. Coincidence?

Russia expropriated oil corporations during its 1917 revolution, when oil prices were high because of World War I. No one argues that high oil prices triggered Russia's revolution. The Soviets nationalized everything, not just oil. If high international oil prices spark resource nationalism, how do we explain Mexico's oil nationalization in 1938 and Iran's in 1951? Neither coincided with high oil prices.[17] The record is mixed in the 1970s. Starting in Libya in 1970, a wave of oil nationalizations began before the international oil price surged from $2.70 a barrel to $9.76 in 1973.[18]

Resource nationalisms increased after oil prices spiked, but momentum had been building. Certainly, high-priced oil made nationalization more attractive to governments. They could collect 100 percent of much higher rents, enabling them to easily buy oil transnationals, set up national oil companies and still pocket huge sums.

National oil companies became dominant in the 1970s, when thirty-eight were set up and nineteen strengthened.[19] This was mainly a trend in the Global South, but Britain launched BritOil and Canada set up Petro-Canada then.[20] High oil prices helped, but the reigning Keynesian ideology, energy security and early environmentalism boosted the trend. The strongest instance of falling oil prices leading to privatization happened in the 1980s and 1990s, when world oil output rose and demand fell. The victories of Conservative Margaret Thatcher in Britain (1979) and Republican Ronald Reagan in the us (1980) sparked international trends to privatization, globalization, deregulation and the defeat of government activism and national sovereignty.

Britoil, France's Total and Petro-Canada were privatized. Norway partly privatized Statoil. The World Bank and the International Monetary Fund pressured Global South states to make their national oil companies act like the private, for-profit ones and abandon social goals. Communism's fall in the Soviet Union and Eastern Europe unleashed a round of privatizations, including those in the oil industry. Privatizations coincided with low oil prices, but only partly depended on them. Both the oil price and the political arguments have merit. The wellsprings of oil nationalisms arise without much regard for oil prices, but price affects the bargaining power of governments. Oil nationalism succeeds better when oil prices and citizens' grievances about foreign oil are highest.

The Renationalizing Trend

Russia and Venezuela separately began the trend back to oil nationalization. Vladimir Putin used state ownership to help restore Russia as a superpower. He moved against Russian oligarchs who plundered Russian resources and made alliances with Western transnationals. Mikhail Khodorkovsky, Russia's richest man, was jailed in 2003 when he was about to sell a controlling stake in Yukos petroleum to ExxonMobil and Chevron. Partnerships with foreign transnationals ended. Re-nationalization should not have surprised the oligarchs. In his 1999 PhD thesis, finished before he became president in 2000, Putin wrote: "Russia's ownership of its strategic resources has critical importance for the country's economic development and strategic global influence.... Oil and gas... serve as a guarantee in Russian foreign affairs.... Only the state, not corporations, shall set long-term strategic priorities for oil and gas development in Russia."[21]

Meanwhile, Venezuela's revolutionary government helped ignite renationalization when it retook control over PDVSA, the state-owned oil company, in 2002–03. Under free-marketeer pressures in the 1990s, PDVSA had acted like a private, for-profit company, and opened service agreements with foreign transnationals. Ordinary Venezuelans, PDVSA's nominal owners, saw few benefits. President Hugo Chavez reversed that and PDVSA funded ambitious programs for the poor. Oil revenues enabled Venezuela to gain policy sovereignty from the International Monetary Fund, the World Bank and America. The impetus to regain control over PDVSA began after Chavez's election victory in 1998. It was four years before oil prices rose, in 2002. Bolivia and Ecuador followed Venezuela's example. Nigeria, Chad, Gabon and big oil consuming Asian countries also caught the resource nationalist wave, but not to pursue revolutionary goals.[22]

US Exceptionalism

Resource nationalisms are usually portrayed negatively in the United States. "Not only are state-run resource companies inefficient compared to private sector firms but also many of them are hostile to US interests," states Irwin

Greenstein, a writer on finance. Barack Obama echoes this view. Foreign resource nationalism "bankrolls dictators," "pays for nuclear proliferation," "funds terrorism" and "makes America dependent on foreign powers," he states. Most American observers see US resource nationalism as good, but don't call it that. Promising that his country would not be held hostage to dwindling resources, hostile regimes and a warming planet, Obama pledged to put America on the path to energy independence.[23]

It's a refrain of all recent presidents. America has a National Energy Policy reminiscent of Pierre Trudeau's NEP. When Democratic and Republican leaders talk about energy independence, they usually mean their own national independence. When speaking to Canadians and Mexicans, though, US leaders refer to "North American" or "continental" energy security. In the latter usage, freedom from "foreign" influences means independence from powers outside North America.

The New Oil Nationalisms

Captive to the "progress" ideology, leading thinkers in every generation since the mid-1800s have incorrectly declared nationalism dead. They believe nationalism will be eclipsed as humanity progresses from the local to the global. But nationalisms, including resource nationalisms, are not declining. They have staying power. Oil and nationalisms make a potent, emotive mix because citizens believe resources should belong to their country and resent foreign corporations getting sweetheart deals. Governments often portray oil as a national salvation and a roadmap to an independent, prosperous future. If you live in a resource satellite, you wait at the end of the line to access your own resources. Look, but don't touch. Colonials get second-rate coffee while the best beans are exported. Rich foreigners drive big cars abroad on your oil, while few locals benefit. Instead, they get the environmental damage. No wonder "colonials," like those in Nigeria's Niger Delta, so resent big, foreign oil.

Few proponents call themselves resource nationalists. Mexico's president Lázaro Cárdenas (1934–40) was the trailblazer. Big Oil got concessions in Mexico that Mexican economist Victor Rodriguez-Padilla calls predatory. "With the complicity of weak or corrupt governments [oil companies] turned the oil areas into regions with their own laws, authorities and police forces," Rodriguez-Padilla wrote. Mexican oil nationalism rose "in reply to the plundering, the arrogance, and the arbitrariness of the concession system."[24] Foreign oil companies treated Mexican workers as peons, refused to negotiate a forty-hour workweek and paid sick time, and defied rulings of Mexico's labour arbitration board, Supreme Court and president. In response, Cárdenas nationalized all oil reserves, facilities and foreign oil companies in one day in 1938. It was immensely popular. "All oil reserves found in Mexican soil belong to the nation," Cárdenas declared.[25] He created Pemex, a government company, to run all aspects of Mexico's oil business. Pemex still dominates, but transnationals have recently gotten service contracts, particularly in exploration. Mexico was the first developing country to take over an entire industry controlled by northern corporations. It's still celebrated as "Fiestas Patrias," or "Expropriation of the Petroleum Industry Day."

Britain, America and the Netherlands, home of big foreign oil in Mexico, boycotted Mexican products but made no concerted effort to topple Cárdenas. The timing was fortunate. War clouds gathered over Europe in the run-up to World War II. Once war broke out, Mexico busted the Western boycott by selling oil to fascist countries. The nationalization stood. Pemex has been run reasonably competently, if corruptly, ever since. It inspired other nationalizations. Iran copied Mexico, but its effort was ill-timed, coming at the height of Cold War hysteria in 1951. Britain and America backed a coup and installed Iran's Shah as dictator, who then reversed the oil nationalization. In 1960, OPEC was formed, linking Venezuela, OPEC's main initiator, with Middle East oil producers. By the 1970s, OPEC was a powerful oil cartel. Most oil-producing countries, including some outside of OPEC, nationalized part or all of their petroleum industry then.[26]

Forty-five colonies won political independence between 1947 and the early 1960s, but economic development didn't happen. The dominant view in the Global South was that the "neo-imperialism" of northern transnationals was the main obstacle.[27] Economic independence was decolonization's second stage. Keynesianism, *dependencia*[28] and socialism, all popular models then, held that the Global South could not develop if foreign corporate control prevailed.

Many citizens in newly liberated countries mobilized around appeals to democracy and nationalism, couched in anti-Western or anti-American language. They inspired similar voices elsewhere, including in Canada. The Waffle movement for an Independent Socialist Canada, the Committee for an Independent Canada, and the NDP urged the takeover of foreign oil and potash companies and campaigned against "corporate welfare bums." These were Canadian ripples in a global wave of 336 takeovers of transnationals in the first half of the 1970s.

In 1974, the United Nations General Assembly recognized each country's right to nationalize resources: "To safeguard its resources, each state is entitled to exercise effective control over them and their exploitation with means…

including the right to nationalize or transfer ownership of such resources to its nationals, this right being an expression of the full permanent sovereignty of the state."[29] The UN set up a Commission on Transnational Corporations and a binding code of conduct for transnationals in 1974. The General Assembly passed in principle a treaty asserting the right of countries to control their resources, limit transnationals' power and expropriate foreign corporations.

An international momentum was building for the sovereign equality of all countries and radical popular sovereignty where rule really was by the people. Oil nationalizations often included social goals like promoting workers' rights, supporting development, paying for public services and gaining a window on the secretive oil business. Conservation was rarely part of national oil company mandates. The paradigm of popular national sovereignty challenged the private, for-profit corporations. If allowed to spread, nationalization would displace the transnationals and diminish the power of major oil-importing countries, including the US and France, which housed the transnationals and partnered with them as extensions of their state policy. Oil transnationals vilified nationalization.

The US vetoed the UN treaty on nationalizing transnationals and set up the G5 in 1975 to include its main economic allies—Britain, France, Germany and Japan—to marginalize and replace the UN. Italy and Canada were soon added, and it became the G7. In 2009, major powers in the Global South, including China and India, joined to create the G20. The G7/20 valorized private corporations and markets over sovereign equality and redistributing the world's wealth. The main capitalist powers had re-established themselves as the world's main players.

Corporations and banks were also alarmed at the de-globalizing wave and founded New Right organizations such as the Trilateral Commission, set up in 1973 by David Rockefeller, Zbigniew Brezinski and others. The Trilateralists came from banks, transnationals, governments, universities, media and conservative unions to create ruling-class partnerships in North America, Western Europe and Japan. Trilateralists decried an "excess of democracy" in which "the democratic spirit is egalitarian, individualistic, populist and impatient with the distinctions of class and rank."[30] Nationalism was the other target. Rockefeller called for "a massive public relations campaign" to explain the necessity for the "withering of the nation-state." These and other efforts ushered in neoliberalism and globalization as the dominant ideology, and claimed the end of national sovereignty was inevitable. A good ideology for corporate rule.[31]

The World Bank, the International Monetary Fund and the World Trade

Organization adopted these ideas in the 1980s and 1990s and presented them as "free trade," "smaller government," "the magic of the marketplace" and "globalization." But opposition grew once people experienced their harshness. The poor in the Global South protested after regulated food prices and protected agriculture ended. But the Global North's media didn't much notice until the dramatic shutdown of the World Trade Organization meetings in Seattle in 1999 punctured the assumption that globalization and corporate rule were inevitable. Opposition also emerged in Russia and Venezuela. Radical, transformative nationalisms, international solidarity and twenty-first-century socialism surged in South America after 2000. Unlike communist revolutions, the ballot box was their route to power.

The current resource nationalisms pick up where the previous versions left off, but with less sovereignty-asserting emphasis. No two nationalisms are alike, but some include environmental and peak oil concerns. New resource nationalisms can still include government ownership, but also public control through partnerships with petro-transnationals, tax and royalty increases, tougher regulations and public interest ownership.[32]

Today's resource nationalisms start from neoliberalism's failures and reassert working for the common good and popular, national sovereignty. Bolivia is an exemplar. Riding a wave of protests against giveaways of Bolivia's natural gas resources, Evo Morales was elected president in 2006. "The neoliberal governments gave away hills, rivers and mining concessions," Morales charged. "We have to start recovering those concessions." After oil transnationals conceded majority control, Morales said, "Bolivia will not be ...a beggar state with many social problems. We will continue in this path of recovering our natural resources . . . that belong to the Bolivian people."[33]

"Nationalization" in Bolivia meant foreign transnationals had to renegotiate contracts and give 51 percent ownership to YPFB, the state oil company. Transnationals get service contracts and joint ventures with YPFB. Bolivia calls this "twenty-first-century nationalization," as an explanation to hide YPFB's shortage of trained Bolivian managers. Resource nationalisms also spread to authoritarian countries like Russia and China.[34] Since an oil shortage would stop China's long march to industrialization, secure access is more important than a low oil price. This is true elsewhere. Thus, national oil companies now search for oil far from home. They compete globally with transnationals.[35]

Chinese corporations moved quietly into the Sands as minority investors in several projects before Beijing-based CNOOC bought up Nexen, a Calgary petroleum company, in 2013.

Can National Oil Companies Pass the Environment Test?

National oil companies are easy to vilify. "Historically, government ownership of private companies has been notorious for lowering productivity, wasting resources, and distorting competition," writes corporate lawyer Simon C.Y. Wong, "often as a result of unclear objectives, political interference, lack of discipline, and poor transparency."[36] That's too much of a blanket condemnation, but Wong points out national oil companies' common negative features. Even if created with the best intentions, they can go badly wrong. They can line the pockets of politicians and corrupt civil servants. Dictators can use them to increase their power. They can be very inefficient.

National oil companies also foul the environment as much as transnationals do. Most are owned by Global South governments, where pressures for environmental protection are often weak. If they exclusively operate domestically, they may get away with practices that would not be allowed if they operated in someone else's backyard. Regional context matters. Norway's Statoil has a much better environmental record than the national oil companies of India, Venezuela and China, but is no role model. It's had sizeable oil spills in the Barents Sea and Arctic waters. It went into an ecologically sensitive area, despite protests by Norwegian environmentalists. Statoil also operates in Alberta's Sands. Why does Statoil work in these areas? Partial privatization ensures that it abandons its public interest mandate and acts like a profit-driven transnational.[37]

In Ecuador, Brazil, Peru, Bolivia and Nigeria, grassroots movements "to keep the oil in the soil" insist that the earth is more valuable if undisturbed. In Nigeria it's a better economic deal, says activist Ivonne Yanez, "to leave the oil in the soil than to sell it internationally, because of the damages," the corruption and all the oil stolen. The movement is strongest in Ecuador. Ecuador hoped to forego development of a 900-million-barrel oil field in Yasuni National Park, part of the Amazon rain forest and South America's most biodiverse area, if northern countries gave Ecuador half the oil reserve's value—$3.6 billion—to use for social and infrastructure goals. Germany, Spain, France, Sweden and Switzerland pledged support, but it was not enough. Ecuador discarded the effort in 2013. Indigenous and ecological groups are fighting to revive it.[38]

Neoclassical economists contend that unless national oil companies ditch non-economic purposes and face the market's bracing winds, they develop negative features. Economists first preference is privatization. Second best is to force national oil companies to act like private, for-profit entities. This is narrow thinking. If a national oil company loses its social or environmental functions, why keep it government-owned? What helpful things can only not-for-profits companies do? If advantages do not outweigh disadvantages, public ownership should not be tried. If, on the other hand, Canada can't get sustainable energy security without not-for-profits, it must start them.

What is the Case for Not-for-Profits in Canada? The following are important tests:

- Will they substantially change the political balance in favour of energy security, climate change action and environmental protection?
- Will they bring sufficient revenues for governments, community groups and First Nations?
- Will they foster more conservation and renewable energy?
- Will they help the transition to a low-carbon society?
- Will they preserve nature and secure Canadians' energy needs in times of global shortages?
- Will they end climate change denial campaigns?
- Will they support a Canada-first, eco-energy security plan? (Or is it in their profit-driven DNA to explore in ever more dangerous areas and sell as much carbon fuel as possible?)

Equally tough questions must be asked of the privately owned oil transnationals in Canada:

Why Not Profit?

Rebranding BP as "Beyond Petroleum," BP spent up to $125 million a year on its corporate social responsibility image, gaining first place in *Fortune* magazine's accountable corporations in 2007.[39] But words are one thing, action another. Driven to save $500,000 a day in rig costs, BP speedily exited from sealing its Deepwater Horizon well in the Gulf of Mexico in 2010. The resulting explosion killed eleven workers, unleashed the biggest US environmental disaster ever and blew the top off BP's public relations efforts. As oil expert Michael Klare remarked, "These managers operated in a corporate culture that favoured productivity and profit over safety and environmental protection."[40] BP was revealed as just another greedy oil corporation.

That's what it should be, according to neoliberalism's founder Milton Friedman.

To Friedman, "social responsibility" means the corporate executive is to make expenditures on reducing pollution beyond the amount that is in the best interests of the corporation. The only social responsibility for business, Friedman adds, is "to use its resources and engage in activities designed to increase its profits."[41] According to Friedman's logic, it's in the nature of corporations to act this way. Corporate managers could try to cut oil output, but corporate law is framed to prevent this, argue David Thompson and Keith Newman in their 2009 Parkland Institute report, *Private Gain or Public Interest.*[42]

Many oil corporations fund climate-change denial. The "denial industry," writes columnist George Monbiot, are "those who are paid to say that manmade global warming isn't happening. The great majority of people who believe this have not been paid: they have been duped....You keep stumbling across familiar phrases and concepts which you can see every day....These memes were planted by public relations companies and hired experts." Simon Fraser University professor Donald Gutstein documents denial campaigns funded by oil, coal and auto companies. "They sow doubt about climate change in Canada and elsewhere," Gutstein writes. By stoking widespread skepticism, denialists thwart strong government action on carbon fuel use. While temperatures rise and glaciers shrink, oil corporations' profits soar. Gutstein documents ExxonMobil's role in backing the climate denial campaign while Lee Raymond headed the company (1993–2005). Shares rose fivefold "when Raymond did everything to prevent action on global warming." Raising climate-change doubts is good for business, but is it in the public or nature's interest to keep a for-profit incentive system that rewards blocking climate action?[43]

All the Economic Rent

A 2007 World Bank report reveals the magnitude of oil rents in Alberta: the average price of a barrel of crude on international markets during 2006 was $65–$75, while the costs per barrel were $3–$5 in the Middle East, $12 in the Gulf of Mexico and $15 in the North Sea. Margins were $50–$70 a barrel. In 2009, Suncor estimated it cost $30–$31 to develop, extract, transport and sell the average Sands barrel. If we add $3 a barrel (10 percent), a "normal rate of profit," total costs were $33–$34 a barrel. Bitumen averaged $47.19 a barrel then. Suncor pocketed somewhere around $16.50 a barrel in unearned rents. Should Suncor get any of nature's capital? It doesn't own it; it didn't put it in the ground. At 440,000 barrels a day, Suncor's unearned windfall profits were around $7 million a day, $2.5 billion a year in 2009.[44]

Alberta pumps 74 percent of Canada's crude oil and equivalents, and 71 percent of its natural gas, so I mainly follow Alberta's story. Saskatchewan and British Columbia matter, too, with the former producing 14 percent of Canada's oil and the latter 22 percent of its natural gas. Their rents are similar to Alberta's.[45] With nine percent of Canada's oil output, Newfoundland's production is falling. It does things differently. Instead of aiming at 100 percent of rents, Alberta Energy's target is 50–75 percent. But it has not even managed that. From 1999 to 2008, Alberta averaged only 47 percent of oil and natural gas rents. If it had hit the target's mid-range, Alberta would have pocketed $37 billion more—and $65 billion more if it had hit the upper range. That's lots of unearned profits to give to ExxonMobil, Shell and other giant petro-corporations.

Albertans and indigenous peoples, the resources' owners, should collect all the rents. The same is true everywhere in Canada. Collecting *all* the rents is a big reason why national oil companies now dominate globally. Public ownership eliminates accounting games and transfer pricing that make corporate profits appear lower so that governments collect less rent. Governments need to collect rent bonanzas now because the resources won't be deliverable for much longer. Like it or not, the world is moving to a lower-carbon energy future. We must have public money to fund the transition. As ecological history scholar John Michael Greer writes, "It took 150 years and some of the biggest investments in history to build the industrial, economic and human infrastructure that turns petroleum from black goo in the ground to the key power source of modern society. To replace all that infrastructure with a new system designed to run on some other form of energy would take roughly the same level of investment."[46]

Unlike the three Western provinces, Newfoundland plays hardball with Big Oil. Conservative premier Danny Williams (2003–10) demanded "super royalties" when the price of oil soars and a government ownership stake in Hebron, an offshore oil play. Chevron refused and demanded $500 million in subsidies. Williams refused in turn. As "companies continue to take home exorbitant profits, it is reasonable and fair to expect increasing returns for the province," he charged. "We are not interested in being held ransom by companies who are unwilling to negotiate fair and reasonable terms.... I would say to those companies that are not interested, to simply move on.... The oil is not going anywhere. And neither are we."[47]

Chevron soon returned and agreed to a 4.9 percent government ownership stake, a super royalty and no provincial subsidy. Newfoundland won similar terms for two other offshore fields. Still, Newfoundland's take is low by Canadian and international standards. Newfoundland's public confrontation contrasts with Alberta and Saskatchewan, but its 4.9–7.8 percent ownership stakes are piddling compared to Norway's two-thirds ownership of Statoil, the world's sixteenth largest oil company.[48]

David Thompson and Keith Newman note that despite favourable public opinion, public ownership has become "an ideological bogeyman of conservatives in the Anglo-American countries. The public ownership no-go zone" has spread to the Liberals and NDP in Canada. How quickly things change. When faced with economic collapse in 2008, right-wing governments threw off their abhorrence of government ownership. If George W. Bush can nationalize much of America's financial sector and buy up General Motors, even as temporary bailouts, anyone can. "Times have changed," Thompson and Newman write. "The ideological, knee-jerk opposition to nationalization clearly has taken a back seat to pragmatic politics and economics. The scale of the nationalization of the global financial sector has been enormous, and unprecedented in its speed. Even conservatives acknowledged publicly that nationalization of financial institutions" was required to save capitalism.[49] Can Canada change course as quickly so the energy industry becomes part of the solution to a low-carbon future?

Doing Things Differently

If Canadians catch today's wave of resource nationalism, will it be different from the last time? No one wants to recreate Petro-Canada. Canadians expected Petro-Canada to transform the industry away from secrecy and gouging of motorists. Transformation went the other way. Petro-Canada became just one more oil corporation. That's what the Conservatives wanted. For them, Petro-Canada stood for big government, state economic intervention, economic nationalism, even anti-Americanism. Brian Mulroney's Conservatives won the 1984 election and ended Petro-Canada's Canadianization mission. It became another arrogant, profit-seeking company. So when the Conservative government began privatizing it in 1990, there was little outcry. When Suncor finally bought Petro-Canada in 2009, the time for mourning had past.

Much is wrong with traditional Crown corporations. Could we do public ownership differently so that control is dispersed among governments at the federal, provincial, territorial, indigenous and local levels? Could much of it be carried out by bottom-up community groups and co-operatives, facilitating conservation, environmental protection, energy security, renewable energy and genuine democracy? Thompson and Newman advocate transforming the oil and natural gas industry in Canada to *public interest* ownership, with novel incorporation rules mandating non-profit entities that wean consumers off carbon fuels and on to renewables: "A public interest industry...would supply the oil and gas that customers wanted, but would not have a profit-maximization mandate. So it would not need to boost sales and consumption or exports, nor would it be compelled to externalize costs. It would no longer be engaged in lobbying and litigation and public relations campaigns to prevent or undermine meaningful conservation or emissions-reduction initiatives.... It would also be a Canadian industry." [50]

Focusing on the pragmatic issue of which type of organization best serves the public good, Thompson and Newman distinguish public interest ownership from public or state ownership:

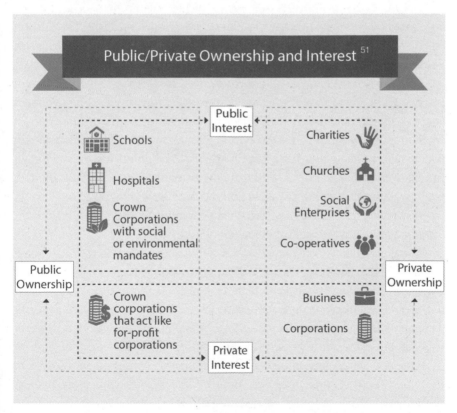

Public/Private Ownership and Interest [51]

The concept may seem politically unfeasible as long as it's cheaper to buy a litre of oil than a litre of bottled water and while the climate remains relatively benign. But when oil shortages and severe climate change disasters strike, many Canadians may change their minds. Crises can open people's minds to ideas they've heard of but rejected as unworkable. The 1970s oil-price shocks, for example, led to hikes in gasoline taxes, fuel-efficient car standards and sweeping oil and gas nationalizations. It's important to circulate good ideas that citizens and governments can pick up when they make sense.

Getting There

Public interest ownership is an attractive idea. The politics of getting there is the tough part. How can petro-corporations' entrenched power and money be overcome? How can workers for these corporations be assured that good jobs will be available after their fall? Thompson and Newman boldly plan to buy the whole Canadian industry in one fell swoop, the way many countries set up national oil companies. "As with all corporate acquisitions, the purchase cost would be paid out of future profits," they write, through long-term, low-interest bonds from revenue generated by future profits.[52] The authors calculate it would cost about $330 billion to buy the whole oil and natural gas industry in Canada. That's a conservative estimate. Whatever the figure, it would be very costly. But as they also note, it's less than the $490 billion the Harper government announced for military spending over twenty years.[53] I asked Keith Newman why they reject regulation rather than purchase. He replied: "The oil and gas industry has immense power due to the income it generates and the jobs it creates. In theory...the objectives we see for the industry could indeed be achieved through regulation, but politically that is not possible due to the power of the industry. So a key objective of public interest ownership is to eliminate private capital from the industry with the intent of reducing its political clout."[54]

Would the plan work? Thompson and Newman assume that the new entities could pay off owners from future earnings, the public interest entities would collect 100 percent of economic rents, and oil and gas rents will remain high. Their plan makes sense only if the carbon-fuel industry continues as usual. The key is for public interest firms to generate enough revenue to promote phasing out of carbon fuels after paying off the old owners.

Professor Mel McMillan, a University of Alberta economist, sees "no sense in buying a firm at a high price that you intend to push into non-existence and without value." He continues: "Acquiring the firm's 'share' of the uncollected rent provides no 'bonus' because that has already been capitalized into the high price that you will be paying. In fact, I suspect it would be very difficult for the public firm to pay off the mortgage. Hence, it is unlikely that those financially burdened firms would have any resources available with which to do 'good [environmental] things' and, indeed, could become a burden on the taxpayers."[55] McMillan is right. If public interest firms use most of their revenue to pay off owners, they'll have little left to encourage conservation and energy security. They'd lose their reason to exist.

Some countries set up national oil companies before buying transnationals' assets. Similarly, smart sequencing could make buying oil corporations in Canada affordable. Bring in most of the Canadian enviro-energy security plan and set up public interest energy firms *before* buying up Big Oil. The eco-energy plan advocated in this book would drive down their anticipated take but also reduce the base for economic rents. They will have to pay the full costs of dumping waste into the air and rivers, which they now do for free. Their stock value will plummet. New public interest firms could buy them for a song and use most of their revenue to promote renewable energy and conservation.

McMillan favours regulation. He argues that governments should collect all the economic rents through royalties, charges, taxes and fees and sell effluent permits that reflect true environmental and other social costs. These measures should substantially raise public revenues and reach environmental goals without ownership costs. However, there's a problem with regulation. Regulation worked well when most economies were nationally oriented, before the era of neoliberal globalization. Canada was an exception, though, because it was already so integrated with the US economy. After 1980, free trade agreements removed restrictions on corporations, enabling them to abandon countries with stringent rules. In the absence of an international carbon-reducing agreement, there are limits to how tough a country can get with transnationals.

Public interest ownership cuts through this weakness because location commitment to Canada could be part of the rules of incorporation. Another problem with only using national regulations is that even if petro-transnationals stay, and ostensibly play by the new rules, they're masters at hiding profits. Low official profits mean low royalty payments to governments. Enron was a notorious case. It hired accounting giant Arthur Andersen to fiddle its books. Outed in 2002, both companies crashed and Enron executives were jailed.[56] Buying oil corporations would not quickly end their political clout, which would continue for some time both inside Canada and via US and British pressure on Canada to reverse course. But as powerful as these obstacles are, they could be overcome if enough Canadians were convinced that we must alter course.

Citizens must be alert to the pressure tactics Big Oil will likely use to try to defeat any Canadian eco-energy plan, whether it's through public interest ownership or regulations, such as issuing a NAFTA challenge. Under NAFTA's chapter 11, corporations can sue Ottawa for lowering their asset value. But Ottawa could counter that its policies are to ensure energy and environmental security, not to punish foreign oil transnationals, and that other countries are adopting similar policies. Better yet, Canada can give six months' notice to exit NAFTA and pass legislation cancelling the older Canada–US Free Trade Agreement. Whether tough regulations or public interest ownership are introduced, oil transnationals will fight hard and dirty. Kevin Rudd, former prime minister of Australia, was turfed by his own Labour MPs in 2010 when the mining industry mounted a multi-million-dollar campaign to discredit his proposed 40 percent tax on "super profits" of coal and iron ore.[57]

If successful at driving Big Oil out of Canada, governments would have to quickly direct the higher economic rents they charge into energy conservation to lower utility bills and create jobs. More jobs are generally gained by conserving a unit of carbon energy than by finding and burning one. It is essential that workers, especially those employed by oil companies, quickly see new jobs open up. It's necessary for Canada to reduce carbon emissions by changing the DNA of energy firms. An eco-energy plan would need strong public support to win a showdown with Big, mainly foreign, Oil. It's crucial to convince most Canadians that climate disruption, local environmental damage and energy insecurity are urgent dangers we can do something about, with public interest ownership as an indispensable tool in a broader plan.

Hybrid Strategy

Rather than fret about buying or regulating petro-corporations, a displacement strategy could combine the strengths of each. If a Canadian eco-energy plan annually cut allowable carbon emissions and provided conservation incentives and revenue streams, public interest firms would flourish. And oil transnationals could stay in place doing upstream production. The transnationals could be told that if they greatly cut carbon emissions, protect watersheds and abide by the eco-energy plan, they can continue—similar to Bolivia's agreements with oil transnationals. But even if the driven-down price for buying petro-transnationals is right, why buy these old dinosaurs made up of staff, machinery and bureaucracies with corporate cultures skilled at finding, developing and selling carbon fuels; ruthlessly pursuing unearned profits; and freely dumping carbon into the biosphere? It would be like buying General

Motors in 2008 instead of letting it die and starting from scratch with new electric-car and hybrid firms.

Selling Both Carbon Fuels and Conservation

The genius of public interest ownership operating within a broader power-down plan is to have the same energy firm sell you both carbon fuels and conservation. Your energy delivery company induces you to buy less of what they sell. This is central to BC Hydro's strategy to achieve 66 percent of its greater capacity needs through conservation by 2020: "This will require building on the 'culture of conservation' that British Columbians have embraced in recent years."[58] Reducing demand will save billions in capital costs and lower users' power bills. It's win-win. BC Hydro and the BC government pay for almost half the cost of energy assessments so homeowners can qualify for a subsidized retrofit. BC Hydro pays residents for their old fridges and picks them up, offers incentives on solar hot water installations and free energy-saving kits for low-income households.[59]

These programs avoid building costly new power capacity. Two conditions enable BC Hydro to promote conservation so effectively. First, it doesn't make sense for competing electrical companies to each have their own transmission systems and power lines. That's what makes electric power a natural monopoly. But without a second condition, a public-ownership monopoly, BC Hydro would face for-profit competitors luring consumers to buy more power from them than from BC Hydro. That's what Alberta's private, deregulated power model leads to—growing power use, skyrocketing rates, higher carbon emissions and competition among distributors.[60]

Natural gas is publicly owned in many countries, but in few countries is conservation a major part of their mandate. Can we adapt BC Hydro's model to natural gas? The key is to have rules so that like BC Hydro, new energy-enviro firms get more net revenue from reducing energy use than from selling more natural gas. The utility model applies to natural gas better than to gasoline. *The Canadian Encyclopedia* describes utilities as "businesses so 'affected with the public interest' that they must be regulated by government regarding entry into...the market, rate charges to customers, rate of return allowed to owners" and to require that they serve all customers in an area.[61] Like electricity distribution, it makes no sense for each gas distribution company to build its own lines in every neighbourhood. But natural gas is not publicly owned the way electric power is in most provinces. Natural gas is either a regulated monopoly with one private, for-profit utility covering a territory or has several competing gas marketers offering consumers fixed charges over a common-delivery infrastructure. Either way, gas marketers have an incentive to convince you to buy more.

Natural gas retailing could be brought into public interest ownership, leaving the finding, developing and pipeline functions in private, for-profit hands. But how could the public interest firms save money, like BC Hydro does on upstream expenses by encouraging less use? One way is through a vertically integrated public interest firm that produces and retails natural gas and benefits from reducing demand and falling exploration and shipping costs. But a problem would remain. Electricity is renewable if it's generated by renewables rather than coal or natural gas. As long as water falls, winds blow and the sun shines, electrical power is generated. Natural gas is different. Without continual exploration, domestic output will fall quickly, and so will revenues for natural gas sellers, whether they're for-profit or public interest firms. If successful at weaning you off natural gas, public interest firms will steadily lose revenue unless they can put you on to the renewables that they also sell.

But that won't be enough. As Richard Heinberg of the Post Carbon Institute shows, conservation will contribute more than renewables to getting to a low-carbon society. A bigger revenue source will have to come from conservation. The best way to incentivize conservation is through tradable national energy quotas (NEQS), an idea pioneered and passed into law in Britain but not yet implemented. It is detailed in chapter 9 of this book and is the best way, when combined with public interest ownership, for Canadians to move to a low-carbon society where everyone can thrive. Each year the carbon energy quota in Canada will decline a little, giving people, governments and businesses time to adjust to lower carbon use and rising prices. Under an NEQ framework, public interest energy firms can get revenue by selling conservation remedies as well as declining amounts of carbon fuel, which will get steadily rising prices. Governments must help, too, by planning better and more public transit, inter-city rail, walkable and denser cities, and safe cycling routes.

What about oil? Oil is easy to transport and can be sold by small retailers without costly infrastructure like electrical or natural gas lines to every customer. But oil's flexibility makes it difficult to create a BC Hydro type of framework. Gasoline sales are not a natural monopoly. How then can we reduce the use of gasoline? Vertical ownership is one way, requiring the monopoly of one firm over producing and selling oil and gasoline. NEQS are a better way. They will raise the price of oil while the energy derived from non-carbon sources such as hydro, wind and solar will not rise, making electric cars and hybrids more attractive.

Canada currently exports over 70 percent of its oil output to the US, and NAFTA locks us into that proportion. If Canadians reduce gasoline usage, NAFTA rules stipulate that most of the oil saved will be available for export to the US so more Americans can drive SUVs and Hummers. That won't convince Canadians to cut back. The eco-energy security strategy proposed in the final chapter of this book creates a positive feedback loop. After exiting from NAFTA and phasing out the Sands and oil exports, Canadian oil use and production will fall, cutting Canada's carbon emissions at the wellhead and the tailpipe. With fewer cars and especially heavy trucks driven, road maintenance costs will drop. Few new roads will need to be built. The savings can be diverted to fund public transit, high-speed inter-city rail, sidewalks in the suburbs and separate cycling lanes. The following questions must be addressed:

- How can we balance maximizing economic rents with "public interest" goals of promoting conservation and securing sufficient energy for all Canadians?

- Would tradable energy quotas provide sufficient revenue for public interests firms to do their conservation work?

- How would governments handle fuel shortages and their impact on transportation, emergency health care and home heating if and when petro-transnationals leave Canada after tough environmental regulations, NEQS and public interest ownership are introduced?

- Can we quickly train enough managers in a non-profit ethos to run new energy firms along public interest lines?

- How can public interest firms balance their revenue needs amidst falling demand for carbon fuels against their environmental mandate?

The goal is reduced carbon energy use, ensuring that every Canadian has sufficient energy supplies and a good living standard.

Public interest ownership is key to resource nationalism. First, private for-profit firms can be domestically owned but can slip into foreign hands. Profit, not nationalism, motivates private ownership. Regulations requiring majority domestic ownership can help, but tend to soften under pressure from powerful owners who can get a better price if sales are open to foreign buyers. Second, governments and citizens—the ultimate owners of the resource—will demand all the rents. Third, publicly owned firms are more likely to act in the public and environmental interest, especially if their terms of incorporation mandate it.

There have been a few tremors for the new resource nationalisms in Canada. Newfoundland won an ownership stake in offshore oil projects, Saskatchewan forced Prime Minister Harper to block the foreign takeover of the Potash Corporation of Saskatchewan, and Harper restricted future control of the Sands by foreign state-owned companies. Anti-colonial and revolutionary resource nationalisms inspire many in the Global South and have some resonance in Canada. But Canada is in the Global North, and environmental nationalism stirs more people here. Australian philosopher Arran E. Gare calls for "an environmentalist nationalism which can harness the legitimate anger against global capitalism to carry out the massive transformations necessary to create an environmentally sustainable civilization." Canadians need to ride this wave.

Chapter 7

Pipelines or Pipe Dreams

It's not a Native thing or a white thing; it's an Indigenous worldview thing. It's a "protect the Earth" thing. For those transfixed on race, you're missing the point. The Idle No More Movement simply wants kids of all colours and ethnicities to have clean drinking water.

—*Idle No More Movement for Dummies*, 2013

Many provinces currently have stronger interconnections in a north-south direction in order to allow for lucrative electricity trade with the United States, rather than in an east-westerly direction that would allow for a pan-Canadian electricity market to emerge.

—Canadian Academy of Engineering, 2009

Pipelines were once thought of as merely ways to convey oil and natural gas. No longer. They now convey controversy. Pipelines ship energy that fuels modern economies. They also ship ecological disasters and violations of indigenous sovereignty and determine who gets and does not get vital oil and natural gas supplies. Opposition is growing to pipelines that encourage the Sands to expand their toxic footprint in northern Alberta and get the dirty oil to tidewater. Opposition to TransCanada's proposed Keystone XL pipeline to carry Sands oil through Nebraska to Texas was so great that in August 2011, two thousand Americans were arrested in front of the White House in an attempt to stop it. That got President Obama's attention. He halted the XL line three months later.

Enbridge's proposed Northern Gateway route to Kitimat on British Columbia's northern coast is wholly within Canada, but has sparked intense opposition. If built, it will take Sands oil to BC's coast mainly for shipment to China. So would twinning Kinder Morgan's Trans Mountain oil pipeline to Vancouver. A third idea, shipping Sands and Bakken shale oil to Montreal and the east coast, has also ignited fierce resistance. Reversing Enbridge Line 9 has been approved. The partial conversion of TransCanada's natural gas mainline, Energy East, to ship oil to Saint John, New Brunswick, is under review. Their combined capacity is much greater than the oil use in Eastern Canada. Energy East is mainly a Sands-exporting line.[1] Because they pass near the burgeoning Bakken shale oil field mainly in North Dakota but also in Saskatchewan, both lines will likely carry American oil to Eastern Canada and for offshore export.

Pipeline routes matter. As long-term commitments to a given route, current and potential pipelines are crucial to Canada's environmental sustainability, democratic sovereignty and energy security. They can determine whether Canadians have the means to get first access to their own energy. They can also lock us into a carbon economy.

All-Canadian Natural Gas Pipeline

Historian William Kilbourn compares pipeline-building to railways in the making of Canada. A line to take natural gas from Alberta to Quebec City (today's TransCanada's gas mainline) caused the Liberal government to fall soon after the Progressive Conservatives, led by fiery Prairie lawyer John Diefenbaker, charged that letting the pipeline fall into American hands was a sellout. "The pipeline debate of 1956 was the stormiest episode in Canadian parliamentary history," Kilbourn writes. "[It raised] most of the classic issues in Canada's survival as a nation: American economic influence and the nature of Canadian-American relations, the debate between north-south continentalism and east-west nationalism; the questions of transportation and national unity, of energy and national growth, of control over natural resources and their exploitation; [and] the latent conflict between western producer and eastern consumer."[2]

Pipeline debates were similar to those about railway routes just after Confederation. Should the Canadian Pacific Railway go the profitable route through Chicago to Canada's Prairies, as British rail magnate Sir Edward Watkin demanded? Or should it follow an all-Canadian route through sparsely populated northern Ontario, as economic nationalists like Prime Minister John A. Macdonald insisted? Macdonald won, ran roughshod over Aboriginal lands in the West, and united Canada by rail and force.

TransCanada's natural gas mainline copied the all-Canadian rail route and became a national icon.[3] Oil picked up Watkin's desired path through Chicago to southern Ontario. These routes have major consequences for Canadians' eco-energy security today. TransCanada's gas mainline bound Canada together and asserted Canadian independence. A shorter line ran from Alberta to BC's coast. Only Atlantic Canada and easternmost Quebec failed to get Western Canadian natural gas. It was a far-sighted project with positive consequences that are still with us. Map 1 shows the TransCanada mainline route that reached Montreal in 1958. A second pipeline route was added a decade later. It shipped Western Canadian gas to Eastern Canadians via the northern states south and west of Lake Superior and Lake Huron.

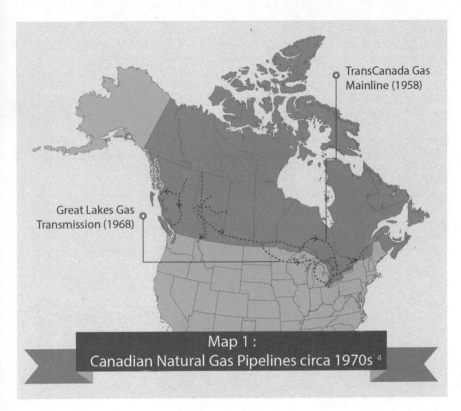

TransCanada Gas Mainline (1958)

Great Lakes Gas Transmission (1968)

Map 1 :
Canadian Natural Gas Pipelines circa 1970s [4]

After 1970, pipelines turned continental, moving Canadian energy to the United States, eclipsing the west-to-east Canadian routes. The Mackenzie Valley Pipeline was a turning point. When Justice Thomas Berger headed an inquiry into the line to bring natural gas from Canada's Arctic to Alberta and the US, and placed a moratorium on it in 1977, he shocked the petro-elites. No longer could they ignore Aboriginal land claims in pipeline decisions. The line was never built, but promoters have periodically tried to revive it.

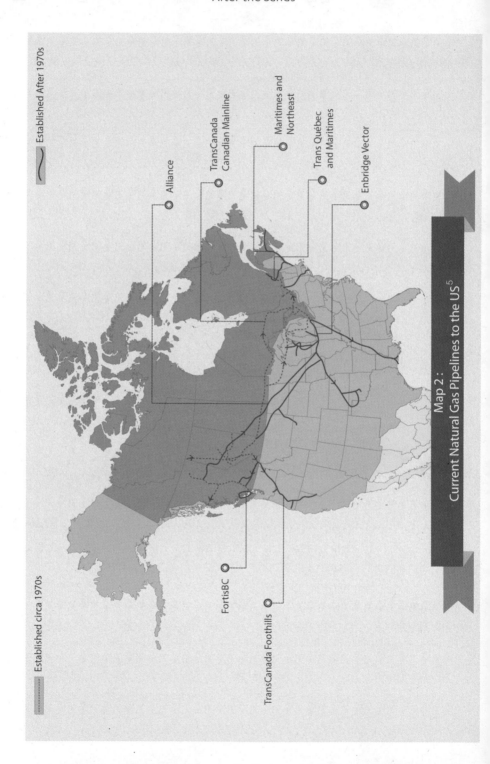

Map 2 :
Current Natural Gas Pipelines to the US[5]

Established After 1970s

Alliance

TransCanada
Canadian Mainline

Maritimes and
Northeast

Trans Québec
and Maritimes

Enbridge Vector

FortisBC

TransCanada Foothills

Established circa 1970s

From National to Continental Pipelines: Natural Gas

The 1989 Canada–us Free Trade Agreement and NAFTA (1994) spurred the building of export lines to America. Map 2 shows the changes after thirty years of continentalism. All natural gas lines built after the 1960s head south. By 2011, almost two-thirds of Canada's natural gas was exported to the us.[6] Meanwhile, Ontario and Quebec import growing supplies of fracked natural gas from nearby states. Exporting natural gas from Western Canada and importing it in Central Canada reduces the use of TransCanada's 14,101-kilometre mainline. It's in trouble. That's why TransCanada wants to convert part of it in order to ship oil to New Brunswick in a retrofit called Energy East. Despite falling volumes, TransCanada's mainline still carries more natural gas, mostly from Alberta, than any other gas pipeline system in North America. It still unites Canadians.

Atlantic Canada has some natural gas. Nova Scotia's Sable Island gas began flowing in 1999. It and some onshore New Brunswick natural gas supply parts of both provinces, but most Maritime gas is exported to Maine's power plants. Newfoundland also has natural gas, but uses it only to assist in oil production. *Globe and Mail* reporter Nathan Vanderklippe asks, "If the east has its own gas and the west wants to send its gas across the Pacific, what's left for the Mainline?" Vanderklippe turns "west" and "east" into persons with wants, when the reality is that big, mainly foreign, petro-corporations create the "wants" in both regions. Canadians' eco-energy needs are not among them. Natural gas exports to the us are falling, being displaced by rising us natural gas output.[7]

No Canadian Oil Pipeline

In the 1950s, oil was treated like just another commodity, not Albertans' birthright or a nation-builder like natural gas. Big Oil and pipeline corporations pushed Ottawa to allow them to sell Western Canadian oil to the huge us Midwest market that sits enticingly between Alberta's oil wells and southern Ontario's market. It was better for them than to build a long, unprofitable route through lightly populated northern Ontario. Canada's de facto oil mainline— Enbridge's Great Lakes system built in the 1950s—brings Alberta oil through the us Midwest before entering Ontario at Sarnia. It was continental, a harbinger of energy-exporting policies forty years later through NAFTA.

Smaller oil independents in Alberta proposed an all-Canadian oil line to Montreal.[8] They lost. Big Oil convinced Liberal Prime Minister Louis St. Laurent that an all-Canadian line was unimportant (see Map 3). Today, we reap the consequences of that poor choice. The us routes make it difficult for Canada to quickly respond to today's challenges of oil insecurity, climate change chaos and the need to protect local environments.

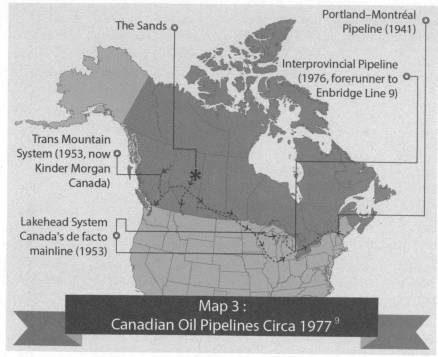

The Sands

Portland–Montréal Pipeline (1941)

Interprovincial Pipeline (1976, forerunner to Enbridge Line 9)

Trans Mountain System (1953, now Kinder Morgan Canada)

Lakehead System Canada's de facto mainline (1953)

Map 3 :
Canadian Oil Pipelines Circa 1977 [9]

Until 2011, oil pipelines from Canada were built to satisfy America's oil addiction. The number and volume of these lines has grown (see Map 4), and with them so has Canada's NAFTA proportionality obligation. Existing oil lines to the US include: [10]

- TCL's Keystone pipeline, to Illinois and Oklahoma—590,000 barrels per day of synthetic Sands crude (not to be confused with TCL's proposed Keystone XL pipeline)
- Enbridge's Alberta Clipper line from Hardisty, Alberta, to Superior, Wisconsin—570,000 barrels per day (ultimate capacity 800,000 barrels per day) [11]
- Enbridge's Southern Lights line—brings diluent from US refineries to the Sands, where it is added to bitumen to move it easily in pipelines
- Houston-based Kinder Morgan's Trans Mountain oil pipeline to Vancouver and Seattle, from Edmonton—300,000 barrels per day, half of it is exported
- TransCanada's proposed Keystone XL pipeline (not approved at the time of writing) would export mainly Sands oil from Alberta to the Texas coast.

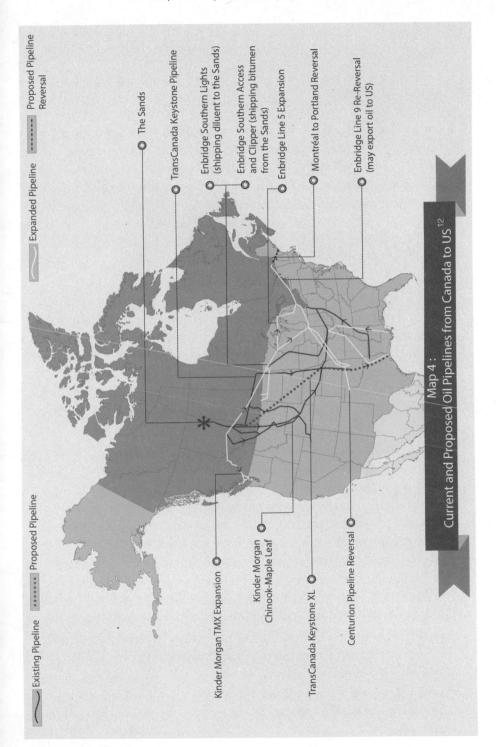

Existing Pipeline

Proposed Pipeline

Expanded Pipeline

Proposed Pipeline Reversal

The Sands

TransCanada Keystone Pipeline

Enbridge Southern Lights (shipping diluent to the Sands)

Enbridge Southern Access and Clipper (shipping bitumen from the Sands)

Enbridge Line 5 Expansion

Montréal to Portland Reversal

Enbridge Line 9 Re-Reversal (may export oil to US)

Kinder Morgan TMX Expansion

Kinder Morgan Chinook-Maple Leaf

TransCanada Keystone XL

Centurion Pipeline Reversal

Map 4 :
Current and Proposed Oil Pipelines from Canada to US[12]

Enbridge's Lakehead system has been Canada's de facto oil mainline to Eastern Canada. It heads east from Alberta and Saskatchewan and enters the US at Gretna, Manitoba. One line goes to Chicago and the other loops through northern Wisconsin and Michigan before both lines meet to re-enter Canada at Sarnia, Ontario. For sixty years, the lines carried Western Canadian oil almost exclusively. Much of it was offloaded in the Midwest states before enough re-entered Canada to supply Ontario refiners.[13] From 1976 to 1999, Enbridge's Lakehead system and Line 9 also delivered Western Canadian oil and energy security to Montrealers.

All that has changed. Passing near North Dakota's Bakken shale oilfield, Enbridge's Lakehead system adds US oil to Western Canadian oil and ships it to Eastern Canadians. When Enbridge Line 9 is re-reversed, it will bring Western Canadian and Bakken shale oil to Quebec via pipeline. It already gets there by train. The tragic explosion at Lac-Mégantic, Quebec, in July 2013 carried US Bakken oil. Enbridge's Lakehead system and Line 9 can no longer deliver energy security to Eastern Canadians. During an international oil shortage, Washington would likely disallow the export of US Bakken oil to Canada. Eastern Canadians need a secure domestic source.

Canadian Energy Self-Reliance Temporarily Won

Despite the route of Canada's de facto oil mainline through the Great Lakes states, Canada gained much "energy self-reliance" during OPEC's rising power in the early 1970s and Iran's 1978–79 revolution. Even though Canada had the capacity to deliver domestic oil to all residents, Eastern Canadians were made vulnerable by John Diefenbaker's 1961 National Oil Policy, which divided Canada into two markets. East of the Ottawa River—Quebec and Atlantic Canada—imported oil. West of the river—Ontario and the West—got Western Canadian oil. Ontarians paid higher than the international oil price for twelve years (1961–1973) to support Alberta's oil industry development.

Subsidization of Alberta's oil lasted a little longer than the subsequent period, in which Ontarians and Montrealers paid a "made-in-Canada" price below the world oil price. This was when OPEC and Big Oil concocted oil shortages to reap windfall revenues. Ontario's earlier subsidy story is conveniently forgotten. Instead, we've heard a selective story about Ottawa and Eastern Canada "robbing" Western Canada of much of its resource wealth through lower Canadian oil prices (1974–1985) before and during the National Energy Program.

Dividing Canada's oil market at the Ottawa River benefited Big Oil. It shortened the distances between their wells in Alberta and their markets in adjacent US states. Diefenbaker's National Oil Policy helped reduce US dependence on Middle East oil by sending Canadian oil to the US Midwest rather than to Eastern Canadians. Big, mainly foreign, Oil conceived the Ottawa River line, and Diefenbaker's Conservatives bought into it.[14] It is still the dominant pattern. It did not and does not make sense to leave almost half of Canadians oil-insecure. There was and is enough domestic conventional oil for all Canadians.

Eastern Canadians experienced the riskiness of importing oil when supplies were cut several times between 1973 and 1979. Pierre Trudeau's government quickly rectified things before starting the National Energy Program in 1980. Ottawa subsidized the building of the Interprovincial Pipeline (now Enbridge Line 9) that brought western oil from Sarnia to Montreal, cutting across the Ottawa River line and displacing much imported oil.[15] The combination of government energy conservation measures, cuts in demand due to rising oil prices, government support for switching consumers from oil to natural gas, and building the Interprovincial Pipeline worked. By 1983, oil imports had fallen to 28 percent of their 1973 level.[16] From 1976 to 1999, Montrealers had secure supplies of domestic oil, demonstrating a major benefit of Quebec remaining in Canada during two sovereignty referendums.

But Canada retreated from energy independence when Brian Mulroney's Conservative government brought in pro–Big Oil policies after 1984. In 1999, without debate or fanfare, Enbridge reversed Line 9's direction, ending Canadian oil supplies to Montreal and bringing substantial oil imports into Ontario for the first time since the 1950s. Enbridge was allowed to reverse Line 9 for "commercial reasons," part of the free market mania unconcerned with Canadians' oil security. Not to worry, the market will always provide.[17] Quebecers now rely on imports for about 90 percent of their oil.

Canada has reverted to its colonial role as "hewers of wood and diggers of oil wells,"[18] buttressed by deregulation and NAFTA's proportionality rule. Re-reversing Line 9 should bring Canadian oil to Montreal again, but will not return us to 1976. Built to make Canadians more energy self-sufficient, Line 9 will now likely bring both Sands oil and US shale oil from North Dakota to Quebecers.[19] Both fuels are more hazardous than conventional oil. Line 9 will help Big Oil but won't give Quebecers sustainable fuel security. More important than Line 9 is TransCanada's Energy East plan to partly convert its natural gas mainline to ship Western Canadian oil east, and add 1,400 kilometres of new pipeline from Quebec to Saint John. If approved and built, it will carry 1.1 million barrels of oil per day, compared to Enbridge's 300,000 barrels per day.

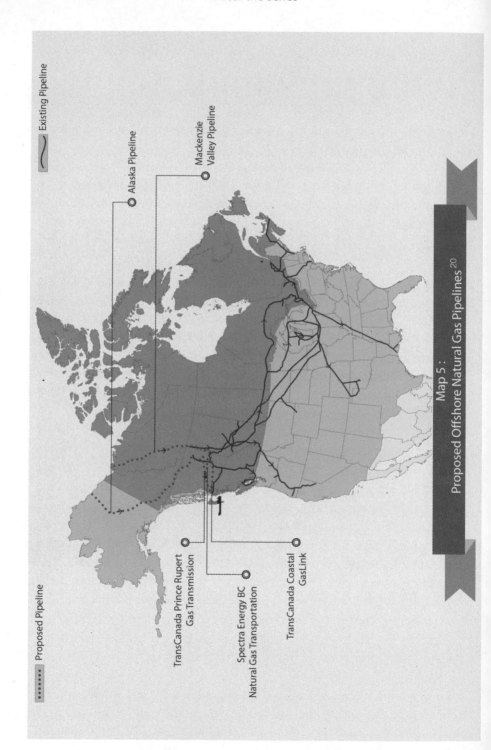

Existing Pipeline

Alaska Pipeline

Mackenzie
Valley Pipeline

Proposed Pipeline

TransCanada Prince Rupert
Gas Transmission

Spectra Energy BC
Natural Gas Transportation

TransCanada Coastal
GasLink

Map 5 :
Proposed Offshore Natural Gas Pipelines [20]

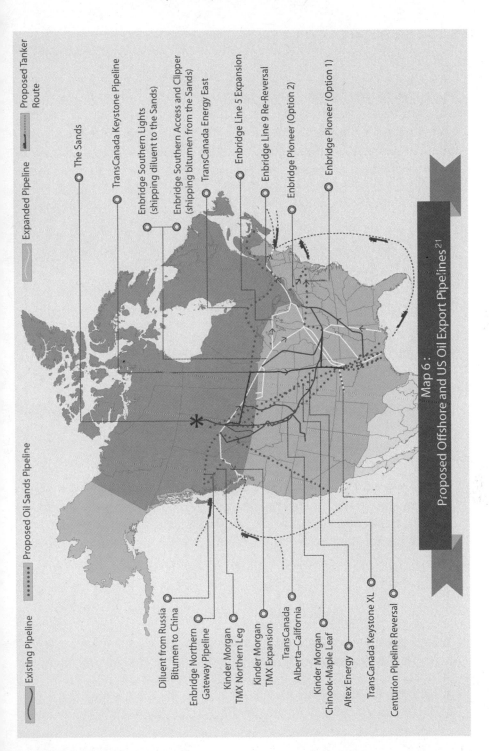

Map 6:
Proposed Offshore and US Oil Export Pipelines[21]

Existing Pipeline

Proposed Oil Sands Pipeline

Expanded Pipeline

Proposed Tanker Route

The Sands

TransCanada Keystone Pipeline

Enbridge Southern Lights (shipping diluent to the Sands)

Enbridge Southern Access and Clipper (shipping bitumen from the Sands)

TransCanada Energy East

Enbridge Line 5 Expansion

Enbridge Line 9 Re-Reversal

Enbridge Pioneer (Option 2)

Enbridge Pioneer (Option 1)

Diluent from Russia Bitumen to China

Enbridge Northern Gateway Pipeline

Kinder Morgan TMX Northern Leg

Kinder Morgan TMX Expansion

TransCanada Alberta–California

Kinder Morgan Chinook-Maple Leaf

Altex Energy

TransCanada Keystone XL

Centurion Pipeline Reversal

From Continental to Offshore Exports

Until recently, virtually none of Canada's oil and natural gas went offshore. Then two events altered everything. US natural gas production surged, reducing imports from Canada, and Obama halted the Keystone XL oil pipeline in 2011. Big Oil in Canada was shocked. They finally did what Peter Lougheed had told them to do six years earlier—open a second front in Asia, where they could get much higher prices for both oil and gas. There are eighteen liquefied natural gas (LNG) proposals to export Canadian natural gas from BC's coast. Only three or four of the proposals are thought to have a good chance of being built.[22]

Although oil is increasingly transported in Canada by rail, barge and Great Lakes tankers, pipelines are and will remain the main way to move large volumes. The following proposed pipelines could bring large amounts of Canadian oil, mainly Sands oil, offshore for the first time (see Map 6):

- Twinning of Kinder Morgan's Trans Mountain oil line—The original line takes Alberta oil to Vancouver and Washington state. Capacity: nearly 600,000 barrels per day, much of which will likely go to Asia.

- Enbridge's proposed Northern Gateway line from near Edmonton to Kitimat, BC. Capacity: 525,000 barrels per day of mainly bitumen, bound mostly for China.

- Partial conversion of TransCanada's natural gas mainline (Energy East) to move western domestic oil (including bitumen) to Saint John, New Brunswick, to be taken by tankers to the US east coast, Texas, Europe or Asia. Capacity: 1.1 million barrels per day.[23]

- Re-reversal of Enbridge Line 9 from Sarnia to Montreal, so Western Canadian oil, including bitumen, gets to Quebec. Capacity: 300,000 barrels per day, most of which was originally to be sent to Portland, Maine, for export, but may now stay in Montreal.

If Northern Gateway is built, bitumen and lighter crude would be taken by tanker from Northern Gateway's terminus at Kitimat through BC's hazardous coastal waters. An informal federal ban on oil-tanker traffic in the straits around Haida Gwaii has been in place since 1972, so would have to be lifted. Strong opposition comes from an alliance of indigenous peoples, environmentalists and trade unionists. Resistance to twinning Kinder Morgan's oil line is also strong. Especially after an oil spill in beautiful English Bay in the heart of Vancouver in 2015, there is widespread opposition throughout BC to the greater tanker traffic that pipeline expansion would attract.[24]

It matters whether pipelines are geared to carry Canadian oil to the US,

Asia or Canada. A pipeline's route can have a decisive influence for the next fifty years. The route can also affect the bite of NAFTA's proportionality rule. The greater the share of Canadian oil going to Asia, the smaller the portion going to the US, dropping Canada's NAFTA export obligation. During an oil shortage in Eastern Canada, Ottawa could divert Canadian oil exports bound for Asia to Eastern Canadians instead, without violating NAFTA. As we've seen, however, Canada cannot send oil to Eastern Canadians quickly, something that is vital in a supply crisis. An equally urgent question is whether Canada should build any new pipelines and continue exporting carbon fuels at all.

Growing Opposition to Pipelines in Canada

Those concerned about Canadian energy security have long demanded the re-reversal of Enbridge Line 9. But when Enbridge's proposal came, many early supporters, including me, opposed it. Why? Wouldn't it revive the dream of giving Quebecers and eastern Ontarians first access to Canadian oil again? No. As environmentalists clearly see, Line 9 is simply a way to move Sands and Bakken shale oil and will not enhance Canadians' sustainable energy security. It will encourage Sands expansion.

Cutting Ontario's access to sweet light crude will make "the province entirely dependent on a single risky source of oil (Alberta's)," wrote Matt Price of Environmental Defence and Gillian McEachern of Forest Ethics. Enbridge's Line 9 re-reversal sparked their *Freedom from Dirty Oil* report.[25] "It is never prudent to rely on a single source of anything," they wrote. Curiously, Gaz Métro and Russia's Gazprom used the no-single-source argument to convince Quebec to approve liquefied natural gas imports.[26]

Imports from OPEC countries and Russia are more secure than Canadian energy sources, Price and McEachern argue. Really? That is far-fetched. Although it makes sense for importing countries to diversify their sources to reduce the risk of being cut off by any one of them, domestically sourced oil is far more secure than any imports. Canadians, though, will win the greatest energy security by transitioning off carbon fuels altogether. All carbon fuels are finite and non-renewable and therefore ultimately insecure. Price and McEachern wield their implausible argument for a good cause—to keep Alberta's "dirty oil" out of Ontario. But must Ontarians or Quebecers choose between Sands oil and less-dirty foreign oil?[27] Why not a third choice—living on Western Canadian conventional (non-fracked) oil?

Enbridge didn't reckon on the surge in environmental and aboriginal opposition. The Sierra Club and Vermont Natural Resources Council got

Vermont's regulator to overturn an earlier approval to reverse the northward flow of the Portland, Maine to Montreal pipeline.[28] It's unclear whether Enbridge will pursue its earlier plan to carry most of the oil, including Sands oil, to Portland, Maine. It would require US approval, including a presidential permit.[29] If not, the oil will stay in Montreal, at least until a higher bidder is found. At first, most will be conventional oil, but Enbridge doesn't rule out carrying more Sands oil.

Ontarians became aware that Enbridge shipped them Sands oil after its line ruptured 3 million litres into the Kalamazoo River in nearby Michigan in 2010. "Learning about Enbridge's poor handling of the rupture, you can't help but think of the Keystone Kops," said Deborah Hersman, chair of the US National Transportation Safety Board. The Board found Enbridge guilty of "pervasive organizational failures."[30] But it was worse. Enbridge knew the line was corroded and cracked, yet did nothing.

Bitumen is hard to clean up. Conventional oil floats, but the bitumen sank to the bottom of the Kalamazoo River. Despite cleanup efforts costing over a billion dollars, the river is still polluted. The Kalamazoo spill was a wake-up call that Sands oil may cause pipeline breaks more readily than conventional oil. In 2013, Sands oil spilled and ran along a suburban street in Mayflower, Arkansas, from an ExxonMobil pipe break, alerting the public to the dangers of piping Sands oil.[31] Massive opposition to the Keystone XL line passing through Nebraska's fragile Ogallala water aquifer led Obama to halt it. Environmentalists oppose Line 9's reversal. Équiterre in Quebec, ForestEthics in Ontario and Environmental Defence contest shipping Sands oil and fracked shale oil on it. Most Quebecers want the line delayed until there is stronger environmental protection. They choose cleaner oil imports over dirty (Sands) domestic oil.

Proposed pipelines in BC and Eastern Canada must cross many lands belonging to indigenous peoples. They've put up the most energetic opposition. To weaken their power, Harper's government passed omnibus Bill C-45 in 2011. Among many other things, it gutted the 1882 Navigable Waters Protection Act. That act, originally meant to assist navigation, was later interpreted to require developers to notify the federal government of plans to cross each waterway, which triggered federal environmental assessments and often extensive consultations with indigenous peoples before construction could occur. Since Canada has so many waterways, development was slowed. No longer. Bill C-45 drastically cut the number of waterways needing assessment.

Annita McPhee, former head of the Tahltan First Nation council in BC, sees Bill C-45 as a "direct attempt to undermine the protection of those lakes and waters and to allow access for developers. This is affecting First Nations, but it's [also] affecting everybody."[32] Bill C-45 and the proposed Sands oil exporting lines across First Nations territories sparked the Idle No More movement in 2012. It quickly spread after Attawapiskat chief Theresa Spence went on a hunger strike. Flash mobs did round dances at malls and city intersections. Roads and railways were blockaded, and sizeable protests spread across Canada and into the US, Sweden, Britain, Germany, New Zealand and Egypt.[33] Many non-natives joined in.

Pipelines aren't the only oil-moving means that are triggering opposition. Rail is making a comeback where pipelines don't exist. Canada's Transportation Safety Board found that an unsafe railway and lax audits from Transport Canada were responsible for the towering fireballs that erupted at Lac-Mégantic in 2013.[34] The train was carrying Bakken shale oil. To its credit, Enbridge had refused to carry Bakken oil with extremely high levels of hydrogen sulphide before the Lac-Mégantic tragedy.[35] But will Enbridge carry shale oil if hydrogen sulphide levels are reduced but not eliminated? The temptation will be strong. Since Enbridge Line 9 no longer exclusively carries Canadian oil, it cannot provide oil self-sufficiency to Canadians. North Dakota shale oil already gets to Quebec. Enbridge will want to carry some shale oil because it's close to their pipelines through North Dakota on its way to Ontario and Quebec.

Energy East to the Rescue?

Should we build the all-Canadian oil pipeline we failed to build in the 1950s? It was a good idea then. Is it a good idea now? Line 9 won't bring Canadians sustainable energy security, but could TransCanada's Energy East oil line do the job? The huge advantage is that it's all on Canadian soil. With a capacity of up to 1.1 million barrels a day,[36] the repurposed line could more than replace all of Eastern Canada's oil imports—about 634,000 barrels per day.

TransCanada's gas mainline was the famous pipeline that in 1957 toppled the Liberal government over Canadian sovereignty. Now it's half-empty. TransCanada's rationale for partly converting it to oil is that it needs higher volumes and that Eastern Canadian oil refiners want access to cheaper Western Canadian oil. Critics charged that TransCanada uses its monopoly power to raise carrying charges on its mainline, driving away Eastern Canadian natural gas customers so it can move more lucrative Sands oil instead. The National Energy Board agreed with the critics, calling TransCanada's request to hike natural gas tolls "cost shifting."[37]

Russ Girling, TransCanada's CEO, portrays Energy East differently, comparing it lyrically to the Canadian Pacific Railway and the Trans-Canada Highway: "Each of these enterprises required innovative thinking and a strong belief that building infrastructure ties our country together, making it stronger and more in control of its own destiny, and this is true of Energy East."[38] Framing it as nation-building is a good way to sell an oil line that will mainly export Sands oil for giant oil transnationals.[39]

Prime ministers are supposed to rise above the fray during pipeline reviews. But Mr. Harper was enthusing in Saint John: "We're not just expanding our markets for our energy projects....We are also at the same time making sure that Canadians themselves benefit from those projects and from that gain energy security."[40] It is rare that Harper strays from the pitch that Canadian oil ensures *North American* energy security. But he did it to promote Energy East. Mike Hudema, Greenpeace's Alberta campaigner, cut through the nation-building rhetoric. "You can't build a nation around a project that will poison water, violate treaty rights and further accelerate a global climate crisis,"[41] he said. East coast consumers want cheaper western oil, but Sands producers want to reach tidewater to get the world oil price. Both hopes cannot be met. Ottawa refuses to intervene. Not only would Atlantic Canadians not get lower gasoline prices, they might not even get Canadian oil. Energy East will sell to the highest bidder.

Harper's government will not order TransCanada's Energy East line to meet all of Atlantic Canadians oil needs before allowing exports. Domestic oil will flow through an all-Canadian pipeline route with native and non-native communities subject to disastrous spills, while citizens get few benefits beyond temporary pipeline construction jobs and a few permanent refinery jobs. But do Atlantic Canadians need to pipe in oil from far-off Alberta? Newfoundland's offshore oil fields produce about enough to meet all of Atlantic Canadians' oil needs. Absurdly, Newfoundlanders can't use the oil they produce. Almost all of it leaves the province and half is exported.

Most Atlantic Canadians live on or near a coast and have no need for a pipeline to bring them oil. It can be shipped in. Despite their risks, tankers must be double hulled and are generally safer than oil pipelines.[42] Further, tankers can be phased out as east coasters' oil use falls. In contrast, an Energy East line would demand Sands oil to fill it for fifty years to pay off capital costs. Newfoundland oil for Atlantic Canadians would follow the precedent of the underwater cable that will bring Labrador electric power to Nova Scotia, strengthening ties among Atlantic Canadians.

If NAFTA's proportionality rule is invoked to stop Atlantic Canadians from getting first access to their own oil, Canada can exit NAFTA after giving six months' notice, something Barack Obama and Hillary Clinton threatened to do in the 2008 Democratic primaries. Obama's halting of TransCanada's proposed Keystone XL pipeline broke the bargain that underlay NAFTA: the US guarantees access to Canadian energy imports in return for the US getting first access to Canada's energy resources. It shows that the US can block Canadian energy exports at will. If so, why should we continue to give the US unlimited access to our energy if that prevents Canadians from using their own oil?

An Oil Pipeline to Stop in Quebec?

If a pipeline carrying Alberta oil to far-off New Brunswick makes no sense, would one that goes to and ends in Quebec? Perhaps. Like everyone else in the world, Ontarians and Quebecers must wean themselves off carbon energy. But even under the national energy quotas plan to do that outlined in chapter 9, they will need secure if steadily diminishing oil supplies for several decades. An all-Canadian oil pipeline makes less sense than it did in the 1950s. As Canadians gradually reduce oil use, the pipeline will become progressively less economic, raising per-unit transport costs and gasoline prices in Eastern Canada.

Ontarians are not overly vulnerable to an oil supply crisis. Less than a tenth of the oil they use comes from offshore, although a growing share is imported and dangerous US shale oil. Ontarians therefore face a moderate energy security risk.

International oil shortages will hit Quebec much harder. It has been importing 90 percent of its oil. There are good prospects for developing oil on Quebec's south shore. It would have to be fracked, though, something that is widely opposed. There are oil prospects in the Gulf of St. Lawrence, too, but they come with great environmental risks. Quebec may not produce much oil soon, if ever. Almost all of Quebec's transportation is fuelled by oil, accounting for more than three-quarters of the province's oil use. Heavy-duty road vehicle use is rising quickly in Quebec.[43] Quebec environmentalists wish to avoid this reality. In 2011, Patrick Bonin, coordinator of AQLPA, a group that campaigns against fracking and on climate change, told me that they focus on leapfrogging to electric vehicles run on Quebec's plentiful hydroelectric power.[44] He agreed, though, that Quebecers will need oil for twenty years or more.

Could a revised Energy East line meet both Quebec's oil security and environmental concerns? If TransCanada converts its gas mainline to oil, would it carry much Sands oil? Chad Friess, a Calgary-based energy analyst, doesn't think so. He expects light oil, much of it Bakken shale oil, will fill the line.[45] After Lac-Mégantic, that won't reassure Quebecers. But unless Energy East is forbidden from carrying bitumen, Energy East will also carry Sands oil. That's why Frank McKenna and Derek Burney, former Canadian ambassadors in Washington, advocated a west-to-east oil line—finding new outlets for Sands oil.

Indigenous peoples and environmentalists wish to block new markets for Sands and fracked oil. Would they still oppose an Energy East line that carried Canadian conventional oil only as part of an effective plan to steadily move Canada to a low-carbon future? TransCanada's natural gas lines are already in place. Repurposing some of them to carry conventional oil is unlikely to disturb the land much more than it already has been disturbed.

Citizens can insist that their provincial government refuse Sands oil. Every province should be persuaded to adopt tough, low-carbon fuel standards like California's. They would effectively keep Sands oil out. After Canadians greatly cut oil use, they won't need any oil pipelines.

The Alternative

Many environmentalists haven't thought through the connection between Canadian energy sovereignty and ecological protection. Conventional oil is less environmentally damaging than Sands, fracked or Arctic oil. We cannot presume that Quebecers and Atlantic Canadians can easily obtain conventional oil from offshore. It's not wise to let for-profit corporations like TransCanada determine whether Eastern Canadians have secure access to domestic conventional oil. Other International Energy Agency countries use governments to ensure oil security. Ottawa should, too.

What about natural gas security? Partly converting TransCanada's gas mainline to oil may mean there won't be enough room for Western natural gas for Ontarians and Quebecers. Deregulation allows more natural gas to enter Ontario and Quebec from the nearby giant Marcellus field in the US, displacing Western Canadian supplies. By allowing TransCanada to convert away from oil, are we trading away Central Canadians' natural gas security?

Relying on Marcellus natural gas is folly. It's fracked, environmentally destructive gas and likely a temporary bubble.[46] When Marcellus declines, and if so much of TransCanada's mainline is shipping oil, how will Ontarians and Quebecers get natural gas? Gaz Métro, Quebec's largest natural gas distributor, opposes Energy East, charging that it will lead to supply shortages and higher prices and threaten Quebec's economic growth.[47]

The proposed 1,400-kilometre new oil line from Quebec to Saint John is attracting protests. So is TransCanada's proposed Energy East line through northern Ontario. The "No Tar Sands Pipeline through Northern Ontario" group contend that "in the fifty-five-year history of this gas mainline there have already been five significant incidents in Northern Ontario. We can expect these numbers to increase as a pipeline built in the 1950s begins use for something other than what it was intended. Oil is much more dangerous than Natural Gas when it leaks into the environment and tar sands oil... especially is nearly impossible to clean up."[48]

Quebec Key to Canadian Oil and Natural Gas Sovereignty

Opposition to receiving Sands oil is strongest in Quebec. Hydro-Québec's phenomenal hydro power encourages Quebec to proudly claim it is meeting Kyoto's targets to limit greenhouse gases. Quebec's dominant political debate has been about whether to stay in Canada. So winning Quebec over to a Canadian oil security plan won't be easy. Its provincial governments, whether federalist or sovereigntist, are put off by the kind of Canadian nationalist rhetoric invoked to support the Energy East line.

Why is Quebec so important? Geography. The four western provinces run on Canadian oil. Ontario gets over 80 percent of its oil from there. Quebec gets more than half of Canada's net oil imports.[49] Atlantic Canada gets a quarter. We saw that Newfoundland can supply all Atlantic Canadians with oil. Ontario takes the rest of Canada's net oil imports. If Canada is going to adopt a Canada-first, eco-energy security strategy, Quebec must be onside. Quebec's acceptance of Western Canadian natural gas matters, too. TransCanada's Energy East conversion plan partly dismantles its natural gas distribution system. If Quebec gets large volumes of Russian liquefied natural gas or US shale gas, it will further fracture that system.

How likely is Quebec to take part? The sovereignty narrative runs counter to the appeal to Canadian sovereignty on energy. The one-third of Quebecers who are sovereigntists have given many electoral victories to the Parti Québecois and Bloc Québecois. Until its 2011 meltdown, the Bloc held the majority of Quebec's federal seats. The Parti Québecois's return to power in Quebec City in 2012 lasted only eighteen months before the federalist provincial Liberals' stunning victory in 2014 on an anti-sovereignty referendum theme. The NDP's 2011 Quebec sweep of federal seats also signals a Quebec open to joining in Canada-wide projects. Quebecers' oil vulnerability is reason enough for their participation in a Canadian eco-energy security strategy. Quebec and Canada have no strategic petroleum reserves. Oil shortages will remind Quebecers of the Montreal ice storm nightmare.

I visited *la belle province* twice recently and got an encouraging reception to the ideas set out in this book. The energy and environmental experts and activists I spoke to told me Quebecers would accept Western Canadian conventional oil. It doesn't matter where Quebec gets its oil from, they said, as long as it's not dirty oil (i.e., Sands oil). A front-page story in *Le Devoir*[50] on February 5, 2008, began a debate on NAFTA's proportionality rule and energy security. It was sparked by a report I wrote for Parkland Institute calling for Canadian strategic petroleum reserves.

Quebecers were shocked to learn that Quebec is extremely vulnerable to international oil supply cuts. The next day, the opposition parties, the Action démocratique du Québec and the Parti Québecois, called for an investigation into Quebec's plans for an oil supply crisis.[51] Louis-Gilles Francoeur, *Le Devoir*'s environmental reporter, then shifted the debate to NAFTA's threat to Quebec's natural gas security.[52] Natural gas meets only 12 percent of Quebec's energy needs yet generates much of its controversies. Gaz Métro had a plan called Rabaska to offload Russian liquefied natural gas (LNG) at an ugly terminal in the St. Lawrence River near the breathtakingly beautiful Montmorency Falls. There was much opposition. Rabaska and another LNG project failed, but not before sparking a debate around NAFTA's proportionality rule and linking it to energy insecurity.

Rabaska's promoters argued that Quebec was vulnerable because it got all its natural gas from Western Canada, a declining source. An alternative supply would enhance Quebec's energy security. Shortly after Rabaska was approved, Francoeur's story argued the opposite—Russian LNG would jeopardize Quebec's natural gas supply because of NAFTA's proportionality rule. If Quebec took Russian rather than Western Canadian gas, the latter would be exported to the US, locking Canada into higher natural gas exports. If Quebec wanted to revert to Western Canadian gas, it would then be unavailable.[53]

The debate around Rabaska LNG shows how ludicrous it is for Canada to export so much natural gas while using the depletion caused by those exports to argue for natural gas imports. Shelving Quebec's LNG import projects strengthened Canada's natural gas system. But soon shale gas deposits were found in Quebec, triggering a debate around whether they should be fracked. Quebec's Liberal government keenly promoted shale gas as a new development opportunity and government revenue source. But fierce opposition forced the government to backtrack, set up a commission to study the issue, and ultimately place a province-wide moratorium on fracking in 2012.[54]

Why is opposition to *le gaz de schiste* (shale gas) so strong? Economic opportunity arguments often trump environmental ones, but not on Quebec's South Shore. Salaberry-de-Valleyfield to Rimouski is Quebec's main farming area. A petition got 128,000 signatures calling for a fracking moratorium.[55] I asked Francoeur why Quebecers so opposed shale gas. First, a 2010 Auditor General's report showed that Quebec mining corporations paid 1.5 percent in taxes, not the 12 percent that the law stipulated. Worse, subsidies and tax breaks were greater than the taxes mining corporations paid. Quebecers expect shale gas corporations to get the same sweet deal.

Second, Quebec's antiquated laws allow mine owners to dig anywhere, including on farms and in towns. When opposition to fracking rose, Quebec's government suggested giving municipalities the right to determine where shale gas corporations can drill. Shale gas corporations demanded huge compensation, making them very unpopular.

Third, many South Shore residents draw water from wells. They heard horror stories from nearby US states about being able to light their tap water on fire after fracking chemicals entered their wells. "Agriculture is part of the culture of the region," contends Lucie Sauvé, a professor at Université du Québec à Montréal.[56] *Gaz de schiste* threatens agriculture. Secret contracts were signed with some landholders. People are insulted that everything is decided secretly. Every regulation is waived for oil and natural gas, she told me.

In 2014, Quebec Premier Philippe Couillard's pro-corporate government extended the fracking ban indefinitely, saying that "social acceptability [for fracking] is not there."[57]

Quebec could import shale gas from nearby US states or avoid fracked gas by depending on Western Canadian natural gas for a generation as it transitions off carbon fuels. It would need to refuse Western Canadian fracked gas. That could work if Canada phases out exports and leaves enough natural gas capacity on TransCanada's mainline. Quebec is very oil-insecure. That could be rectified quickly with Canadian conventional oil and conservation. Hydroelectricity is Quebec's saving grace. It will continue long after the age of carbon fuels ends.

Can Newfoundland Energize All of Atlantic Canada?

Atlantic Canadians should be among the world's most energy-secure people. Combined, they have enough oil, sizeable amounts of natural gas and enough hydro to supply all their current demand for electricity. The four provinces have sufficient potential to satisfy rising electrical power needs when they transition off carbon fuels. But none of these things is happening. A region rich in energy resources is poor by policy. It's the usual culprits—an exporting fetish and each province going its own way. Nova Scotia imports 80 to 85 percent of all the carbon energy it uses,[58] which is typical of the region. All Atlantic Canadians are very energy-vulnerable, ironically none more so than Newfoundlanders and Labradorians, who can't access much of their own oil because their only refinery can't process most of it.

Disputes between provinces block the shift to self-sufficiency. In 1966, Newfoundland Premier Joey Smallwood signed an outrageously one-sided deal to send hydroelectricity from upper Churchill Falls in Labrador to Quebec for seventy-five years. Newfoundland gets a quarter of a cent a kilowatt, an extremely low price in the 1960s. It's almost free today. Incredibly, the contract has no inflation escalator clause.[59] The price will fall to a fifth of a cent in 2016. Hydro-Québec exports most of the power and gets $1.7 billion a year from Churchill Falls, while Newfoundland gets $63 million.[60] Churchill Falls now has North America's second-largest hydroelectric generating capacity, but Newfoundland must send most of it to Quebec until 2041.

The heating oil shortage that briefly hit Cape Breton Islanders in December 2007 showed Atlantic Canadians' vulnerability. About half still use fuel oil to partly or fully heat their homes. Gasoline supplies are also at risk. The Halifax area faced shortages in 2011 when lightning struck Imperial Oil's Dartmouth refinery. It "would make sense to have a supply of gasoline on hand should the unthinkable occur," the *Amherst Daily News* noted. "Imagine if something major were to occur causing those short-term gas shortages to be more frequent and longer lasting?"[61] Indeed. Why are there no plans for such occurrences? The US has a heating oil reserve in its northeast for such eventualities, in addition to huge strategic petroleum reserves on the Gulf coast—for Americans only. Strategic petroleum reserves are national. Oil-sharing agreements in the International Energy Agency are unenforceable. Washington acts on national energy security. Canadian governments rarely think about it except to use America's own security obsession as a pretext for sending them more Sands oil.

It doesn't have to be this way. Newfoundland's oil output, 197,621 barrels per day in 2012, has fallen by half from its 2007 peak, but is about as much as Atlantic Canadians use.[62] Declining production will likely slow but not reverse as output rises in the White Rose field and Hebron's heavy oil field starts in 2017.[63] Oil imports and exports should end. Despite the long-run decline, Newfoundland's oil output is sufficient to meet Atlantic Canadians' needs if the region is part of a Canada-wide conservation plan, set out in chapter 9 of this book. The Atlantic provinces, with help from Ottawa, should start programs to transition off heating oil and better insulate homes and buildings. By cutting oil use faster than Newfoundland's oil output falls, east coasters can be energy-secure as they get to a low-carbon future.

Instead, a great deal of Newfoundland oil is exported. Virtually all Newfoundland oil is refined out of province. The refinery at Come By Chance (also known as North Atlantic Refinery) was geared for heavy foreign oil before Newfoundland began producing light oil. The refinery still is, but now processes a small amount of oil from White Rose in offshore Newfoundland.[64] It's impossible to tell exactly how much Newfoundland oil stays in Canada and how much is exported. Transparency is lacking. Statistics show where refiners get their oil, not where final consumers get theirs. We know that about 53 percent of Atlantic Canada's oil, most of it produced in Newfoundland, goes to foreign refiners.[65] The Irving refinery in New Brunswick can process Newfoundland oil but is geared mainly to exporting gasoline and other refined oil products to the US. Over two-thirds of oil refined in Atlantic Canada is exported.[66]

Atlantic Canada has less potential to be self-sufficient in natural gas, but has important existing and potential sources that, if reserved for Atlantic Canadians, can considerably enhance their energy security. Sable Island, the region's main natural gas field, began producing in 1999. Most of its output is exported to New England. In 2010, Imperial Oil, the lead owner of the Nova Scotia field, deemed it not economical to expand Sable production to counteract declines.[67] But Nova Scotia has other natural gas prospects offshore.

Newfoundland and Nova Scotia produce 4.3 percent of Canada's natural gas. That sounds minor, but Atlantic Canada has only 6.8 percent of Canadians and uses less gas per person than average. No extensive natural gas network reaches them. Besides, Canada produces more natural gas than Canadians use. Thus, Atlantic Canada can achieve natural gas self-sufficiency if it supplies the region and gradually reduces use rather than exporting the gas until it runs out. Newfoundland and Labrador has enough hydroelectricity to meet

95 percent of power needs in all four Atlantic provinces. If hydro power in New Brunswick and Nova Scotia are added, hydro output slightly exceeds the region's total use. Moreover, Newfoundland plans to develop lower Churchill River, one of the continent's major untapped hydro power sites. If earmarked exclusively for the region, it could meet all future expanded power needs for the medium term.[68]

It's not simple. Hydro causes ecological problems. It sounds green and largely is—after major initial damage is inflicted. Damming rivers and flooding major areas swallows up whole ecosystems, often depriving indigenous peoples of their land. Rotting organic matter releases from flooded areas a lot of carbon, and eventually methane. Dams block migratory river animals like salmon and trout.[69] However, using hydro dams already in place is not like building new ones. The latter is not a good idea. Some environmentalists advocate using river turbines to allow animals and rivers to flow normally. River turbines don't allow much water storage for release during peak demand to counterbalance times when winds don't blow and the sun doesn't shine. Others are opposed because water diversions to turbines impact fish stocks. Companies running such turbines have frequently broken the rules in BC.

Instead of using Newfoundland's current hydro power to satisfy Atlantic Canadians' needs, electricity is produced mainly by dirty steam and combustion methods (71 percent) in Prince Edward Island, New Brunswick and Nova Scotia. Nuclear power accounts for 15 percent of New Brunswick's electricity, leaving risks from radioactive waste and reactor meltdowns.[70] Atlantic Canada could instead rely mainly on Newfoundland hydro and new renewables, and become a truly green power generator.

Getting electricity from Newfoundland to the Maritimes is complex. The best route is through Quebec for transmission to New Brunswick, Nova Scotia and Prince Edward Island. But because of bad blood with Quebec over the 1966 Churchill Falls deal, Atlantic premiers agreed to cable routes in order to bypass Quebec. Labrador power will cross the Strait of Belle Isle to join Newfoundland's transmission system. From there, a 180-kilometre undersea cable will bring it to Nova Scotia, at a cost of $1.2 billion. But most of the power won't stay in the Maritimes. Nalcor, Newfoundland's public hydro company, plans to sell it to New England states for a higher price.[71] Two obstacles to sustainable energy security for the region—inter-provincial rivalries and the fixation on resource-exporting—must be overcome.

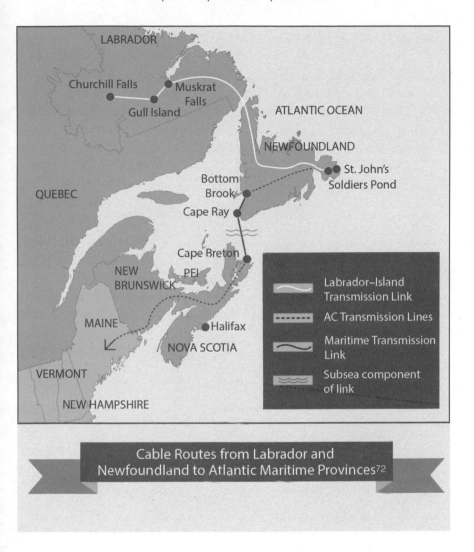

Cable Routes from Labrador and Newfoundland to Atlantic Maritime Provinces[72]

The Vital Questions

We saw that debates over oil and gas pipeline routes in the 1950s raised national sovereignty and national unity issues. They will again. But environmental and indigenous sovereignty claims are now central in ways they weren't then. All these issues will be in contention as Canadians debate pipelines to ship Sands oil to BC's coast, Texas refineries, and Ontario, Quebec and New Brunswick. The vital questions are how to drastically reduce our assault on nature and whether Canada's dwindling conventional carbon resources should meet the energy security needs of others or of Canadians. Canada has all the oil, natural gas and coal needed to transition to a low-carbon future.

Remarkable hydro power—Canada has 12 percent of the world's total—gives us a large base of renewable energy to add new renewables to—wind, solar, geothermal and biomass. Few countries have these options in these quantities. But we live in a vast, cold country, so we need large renewable energy sources. We should make the most of them. Canada will not do so, though, until it shifts direction and breaks the petro-elites' political power. It can be done. Coal used to be king—it is no longer.

Chapter 8
Let Goods Be Homespun

First they ignore you, then they laugh at you, then they fight you, then you win.

—Mahatma Gandhi

These [carbon] numbers make clear that with the fossil fuel industry, wrecking the planet is their business model. It's what they do.

—Naomi Klein, 2012

Ideas are international. By showing what's possible, victories in one part of the world give people hope in others. International agreements on reducing carbon emissions are crucial, but only national and local governments can take the necessary steps to save local habitats and humanity's common biosphere. Led by Dr. Catherine Potvin, a biology professor at McGill University, sixty Canadian scholars from every province have forged ties across provincial, linguistic and disciplinary divides by creating a national plan to get Canada decarbonized. Their 2015 *Acting on Climate Change* report proposes a national carbon tax or cap-and-trade system, and an end to subsidies for carbon energy companies.

The scholars urge Ottawa to terminate oil and gas exemptions from its plan to reduce emissions. They also propose changing urban and rural design to bring people closer to amenities and raise densities, with the goal of aiding decarbonizing.[1] It's good to see the scholars advocating east-west electrical connections to enable provinces with hydroelectricity surpluses sell them to their neighbours. The action plan also proposes energy efficiency gains, especially in buildings, and a revolution to electrify transportation and promote collective and active transport (walking, cycling and public transit).

Citizens' initiatives are helping to lead the way to a low-carbon Canada. The Council of Canadians, Greenpeace, the Suzuki Foundation and many other citizens' organizations are taking action to stop the export of Sands oil and to promote renewables and conservation. Their message is resonating with more and more Canadians, especially the young. Maude Barlow and her team of well-informed activists at the Council of Canadians have held big meetings in just about every community along TransCanada's Energy East pipeline route, from Saskatchewan to New Brunswick, to explain that the pipeline will carry Sands oil and endanger their local habitats. TransCanada is running scared, and has hired Edelman, the world's largest public relations firm, noted for doing dirty tricks.

In Quebec, a coalition that includes STOP oléoduc and Alerte Pétrole Rive-Sud and is supported by Équiterre, Nature Québec and Greenpeace is working to block Sands oil from traversing the province. Gabriel Nadeau-Dubois, fiery leader of the massive 2012 Quebec student strikes, announced on television in 2014 that he is giving his $25,000 prize money for the Governor General's Literary Award to a Quebec organization campaigning against piping dirty Sands oil across Canada and through Quebec. He urged everyone to join in and double his donation. Instead, donations quickly rose twelvefold, to reach $300,000. Nadeau-Dubois's action struck a chord because two-thirds of Quebecers oppose the Energy East pipeline.[2]

For the first time, indigenous peoples are asking settlers for their help in environmental struggles. The battle against fracking in New Brunswick is an example. Martin Lukacs wrote in the *Guardian* that finally honouring indigenous rights is Canada's "best chance to save entire territories from endless extraction and destruction. In no small way, the actions of Indigenous peoples—and the decision of Canadians to stand alongside them—will determine the fate of the planet."[3]

Back to the Future

In 1932, Bertrand Russell wrote, "It will be said that, while a little leisure is pleasant, men would not know how to fill their days if they had only four hours of work out of the twenty-four.... It is a condemnation of our civilization; it would not have been true at any earlier period. There was formerly a capacity for light-heartedness and play which has been to some extent inhibited by the cult of efficiency."[4] Efficiency supplanted sufficiency, and with it the time and inclination to enjoy what one has.

Several authors urge a return to sufficiency.[5] Sufficiency is having an optimal level of material goods that everyone needs to live "the good life." It means "a sense of 'enoughness' and 'too muchness,' a quality where concern for excess is paramount in the life of an individual, an organization" or a nation.[6] Sufficiency recognizes that continually striving for more will not bring greater happiness but triggers destructive climactic forces that will make us very unhappy or dead. Gandhi said it well: "The world has enough for everyone's need, but not enough for everyone's greed." Most of the world's great philosophies treat money as a means to enjoyment, not an end in itself.

In their thoughtful book *How Much Is Enough?*, Robert and Edward Skidelsky write that "an economic order subservient to human ends—one in which riches exist for man, not man for riches...remains the goal of Catholic economic thought."[7]

Similar ideals have been manifest in India, China and Ancient Rome. Despite variations in belief, placing limits on greed has been seen as necessary to attain the good life in almost all civilizations. It cuts across the political spectrum, from Edmund Burke to John Maynard Keynes to Karl Marx. Burke, widely considered the father of modern conservatism, advocated frugality as "founded on the principle that all riches have limits."[8] "The great Error of our Nature, is not to know where to stop; not to be satisfied with any reasonable Acquirement, not to compound our Condition; but to lose all we have gained by an insatiable Pursuit after more."[9] Marx followed Hegel in believing people require only a "sufficiency" of material goods. In the future communist society, Marx foresaw no set limit to what people would have, but assumed that after escaping exploited and alienated labour, they would choose creative leisure over working long hours to make ever more things.

Like Marx, Keynes thought we had to go through capitalism before reaching the ideal society. "We shall once more value ends above means and prefer the good to the useful," he wrote. "We shall honour those who can teach us how to pluck the hour and the day virtuously and well, the delightful people who are capable of taking direct enjoyment in things, the lilies of the field, who toil not, neither do they spin."[10] John Stuart Mill contended that "the best state for human nature is that in which, while no one is poor, no one desires to be richer, nor has any reason to fear from being thrust back, by the efforts of others to push themselves forward."[11] This spirit animated Occupy Wall Street protesters in Zuccotti Park.

Sufficiency lived on among leading thinkers until after World War II, when workers' wages rose above subsistence levels. Advertising, the spread of consumerism and Fordism, the idea that workers are paid enough to buy the things they make, all took hold. Capitalism now depends on endless wants. "That is why, for all its success," write the Skidelskys, capitalism "remains so unloved. It has given us wealth beyond measure, but has taken away the chief benefit of wealth: the consciousness of having enough."[12]

Economists conceived an insatiable economy by erasing the distinction between needs and wants. They've taught millions that there's no such thing as the good life, only a range of desired lifestyles. Since one cannot want too much, one never has enough.[13] University of Michigan professor Thomas Princen explains how, consequently, *efficiency* rather than sufficiency became dominant: "Efficiency is now so universal, so internalized by nearly everyone... that one hardly thinks about it, let alone questions it. The incentives and pressures around efficiency come from all directions. They are ubiquitous, just the way things are."[14]

Efficiency was originally used in Aristotle's sense of effectiveness, until Alfred Marshall, a father of neoclassical economics, reframed it as business managers' skills to organize capital and labour to produce the greatest output with the least input in the interest of maximizing profit and growth.[15] Wendell Berry laments how efficiency requires "a relentless subjection" of means to immediate ends: "Instead of asking a man what he can do well, it asks him what he can do fast and cheap." Today, efficiency means produce more, consume more. Faster and better.[16]

But as productivity grows in high-income countries, Washington-based economist David Rosnick shows, there can be a trade-off between higher consumption and working less, with implications for greenhouse gas emissions. In the early 1970s, Americans and Western Europeans worked about the same number of hours a year, but by 2005 Europeans worked half as much because of longer vacations, lower labour-force participation, earlier retirement and shorter workweeks. Working less means consuming less and emitting fewer greenhouse gases.[17] If Americans worked European hours, Rosnick calculates, American additions to global warming would fall by a quarter to a half.

Why does working more produce greater greenhouse gases? Boston College sociologist Juliet Schor explains: "When households spend more time earning money, they compensate in part by purchasing more goods and services, and buying them at later stages of processing (e.g. more prepared foods). People who have more time at home and less at work can engage in slower, less resource-intensive activities. They can hang their clothing on the line ....They can switch to less energy-intensive but more time-consuming modes of transport (mass transit or carpool versus private auto, train versus airplane). They can garden and cook at home. They can meet more of their basic needs by making, fixing, doing, and providing things themselves."[18] Americans worked less after the 2008–09 recession. It's partly a cultural shift and partly out of necessity. Will Americans continue to downshift if economic growth and more jobs return?

Demand Less

We can't afford to follow a neoliberal economy based on excess. It won't be easy, though, for people to forego luxuries many perceive as their birthright. People are no happier, but material expectations are much higher now than they were on the working-class street in Toronto where I grew up in the 1950s. People will condemn consuming less energy and fewer products as an unjust imposition. We'll see plenty of infantile refusals to face the new reality.

"The campaign against climate change is unlike almost all the public protests that have preceded it," George Monbiot writes. "It is a campaign not for abundance but for austerity. It is a campaign not for more freedom but for less. Strangest of all, it is a campaign not just against other people, but against ourselves." [19] As we lower material consumption, it's important to make a huge cultural shift to remaking and thinking of work as intrinsically rewarding rather than assuming its main reward is a paycheque that enables you to relax after work by buying stuff and passively consuming entertainment.

It's not only the political right and centre that need to switch to sufficiency. The political left and unions must, too. Theirs is a history of fighting for better pay. The unstated assumption is that workers and most citizens don't have enough. That was mostly true in the Global North until the 1950s. A sizeable minority still don't have enough, but the majority do. Instead of more stuff, the latter need fewer work hours, good working conditions and economic security and they need to become active creators in their work, their communities and the regeneration of local environments. A few embrace ascetic lifestyles and poverty vows so they can contemplate creation, beauty and the sublime. Think of monks, nuns and starving artists. But most people are not drawn to self-denial. How can they adjust to a society of sufficiency?

People accepted it in World War II. With slogans like "use your cook stove to cook Hitler's goose" and "put your family on the Victory diet," Ottawa convinced Canadians that rationing helped the war effort. Would such appeals work in today's affluent, individualistic times? Do we need a common enemy or a wartime sense of crisis to accept limited energy use? The response in Japan after a giant tsunami struck the Sendai coast in 2011, killing 15,884, shows how ordinary people can react to peacetime emergencies. Japan closed its fifty-four nuclear reactors and lost a quarter of its electric power.[20] Blackouts never happened because people followed the government's request to reset temperatures to twenty-eight degrees Celsius in homes, offices and stores, and turn appliances and electronic devices to low-energy settings. They shut off lights and escalators. Power demand dropped 15 percent the first summer after Fukushima, and a still-impressive 10 percent the next summer. People conserved for the common good.

Although ordinary Japanese citizens had to adjust, they overwhelmingly preferred conservation to restarting nuclear plants. Must people outside Japan experience a similar crisis before demanding that their governments initiate major cuts carbon-energy use? No one wishes a repeat of Japan's tragic loss of life. But we should stop doing reckless things like building nuclear reactors beside oceans to cool them with seawater so we can have excessive power. The best way to avoid another Fukushima is to seriously conserve now.

How Much Is Sufficient?

Boosting renewables and finding energy efficiencies isn't enough. Like it or not, a conserver society is coming. Will we direct it to maintain equity, energy security, social cohesion and enjoyable ways of life? Or will we let the market take us where it will? Not acting chooses the latter by default. It could lead chaotically to Thomas Hobbes's dystopia, in which life is "nasty, brutish and short." The success of citizens' movements such as food sovereignty and Transition Towns show the shift to a conserver paradigm is occurring. Four principles can guide us:

- Sufficiency to replace efficiency as the central economic goal
- The move to using less energy based on equitable sharing; energy as a human right
- Co-operation and de-globalization to replace the greed of each as economic guides
- Loosely linked national and local economies run in bottom-up democracies to replace a single global, corporate economy

These principles may be fine for armchair philosophers. Will real people want to or be able to follow them? Will they make life better? Efficiency is central to the progress story. The succession of wondrous new technologies based on cheap carbon fuels leads many to believe that we can escape scarcity and that growth will continue forever. The progress faith goes like this: When easy oil ends, science will surely invent substitutes and we will continue our upward climb. Human ingenuity almost replaces the Creator among the progress flock.

More now doubt the progress story. In 1972, the Club of Rome think-tank cast doubt with its million-selling book *The Limits to Growth*. The MIT authors showed that "continued growth in the global economy would lead to planetary limits being exceeded sometime in the twenty-first century." Population and economic collapse is likely.[21] Although the authors said collapse is not inevitable, mainstream opinion dismissed the book with that charge. Forty years on, many accept that we live on a small, finite, lonely planet. David Suzuki compares continual growth in a finite world to "the creed of a cancer cell."[22] Many non-renewable resources are being depleted. Book titles containing "The End of" are in vogue. End of growth, end of food, end of nature. While they over-dramatize, they reflect an awareness that doubts endless progress and sees human activity leading to widespread destruction. Environmental economist Herman Daly developed a steady-state economy model based on the following principles:[23]

- Renewable resources must be used no faster than the rate at which they regenerate.

- Non-renewable resources such as minerals and carbon fuels must be used no faster than renewable substitutes can replace them.

- Pollution and wastes must be emitted no faster than natural systems can absorb or recycle them.[24]

Living by Daly's rules means burning much less of the carbon fuels that enabled unlimited global travel and transportation. They won't stop, but will greatly diminish, ending globalization in the sense of the death of distance. But a conserver society should still be able to support the best of globalization—rapidly advancing communications that connect the globe and shrink the psychic, social and cultural distances among people. Daly is right, but his "steady-state economy" implies stagnation. We need to grow the good and shrink the bad. Jean Lambert, a British Green Party member of the European Parliament, wants to shift the debate. "You have the growth of children, which we all think is very important and very positive, and then you have the growth of cancer, which we all think is very negative." They're both growth. What do you want and not want to grow?[25]

Stories of coming climate change and peak everything are part of current culture. Instead of hoping great new inventions will allow us to carry on as usual, I explore the cultural turning we need to make, adjust to and thrive under the limits we face. It is more a returning than a turning to sufficiency.

That ethic guided us through most of human history. By sufficiency standards, the majority in the Global North have enough to live the good life. Today, the ethics of frugality and prudence that once guided most societies are seen as stultifying. Few realize that only the abundance of cheap oil has allowed us to chuck out thrift along with hot water bottles. Must a thrifty society be joyless and puritanical? Only if we insist on pursuing insatiable wants and fruitlessly working long hours to fill constant yearning for material baubles. We need to learn how to live on less stuff and do leisure again. Once the basic needs of food, shelter and clothing are met, most of us get our greatest pleasures from valuing each other and nature.

Energy as a Human Right

Ten thousand years ago, writes Andrew Nikiforuk, "a hunter-gatherer collected the equivalent of 1.5 barrels of oil a year from plants and animals. ... But by 1880, coal and steam slaves had exploded the amount of energy available to the average person to the equivalent of fifteen barrels of oil .... A hundred years later, Europeans gorged on twenty-six barrels per person annually." Americans supergorged on twice as much—fifty barrels per capita.[26] Canadians were in between. Pre-industrial revolution people couldn't support what we now consider essential to developing everyone's capacities—freedom from endless toil, access to universal public health care, and time for leisure and education.

Canadians love to complain about outrageous gasoline prices. The reality is they're way too low. Even after oil spiked to $147 a barrel, a litre of gasoline in Canada was still cheaper than a litre of free city water bottled and sold as "nature's own." What other liquid sells for a dollar and change a litre? That litre replaces four-and-a-half weeks of human labour. When oil prices spike again, it could cost five dollars. That may still be a bargain, but many Canadians wouldn't be able to afford it. Access to energy is a necessity. Ever since humans discovered fire, we've sought fuels with which to cook food and stay warm. The latter is especially important in cold countries. In earlier times, gathering firewood or cow dung and keeping draft animals to do heavy lifting met most energy needs. Now more than half of us live in cities where such energy sources are rare or non-existent.

Current energy needs go far beyond traditional ones. Getting an education, health care and citizen participation all require expenditures of energy. Lighting is crucial for reading and education. Women are often denied opportunities because of the traditional, time-consuming burdens of collecting water and fuel wood and cooking without labour-saving appliances. Street lighting enhances safety and the right to pursue educational, employment and social activities in the evenings. Internet and computer use are key to getting work. All require electricity. Space conditioning, defined in India as keeping homes at comfortable temperatures, is another widely recognized right.[27]

Those without basic energy are energy-poor, defined by the United Nations as "lack of choice in accessing adequate, affordable, reliable, high quality, safe and environmentally benign energy services to support economic and human development."[28] In 2000, oil was very cheap. Even so, almost a third of humans, and 99 percent in the Global South, were deemed energy-poor.[29] Like access to safe water and sanitation and the right to a healthy environment, energy should be recognized as a human right.

Campaigning around energy as a human right can be effective. Once their government signs on, citizens can hold it to account. But it must be framed around transitioning off carbon fuels. Otherwise, Big Oil will try to co-opt the right to sell their lethal products to a new market—the world's poor.

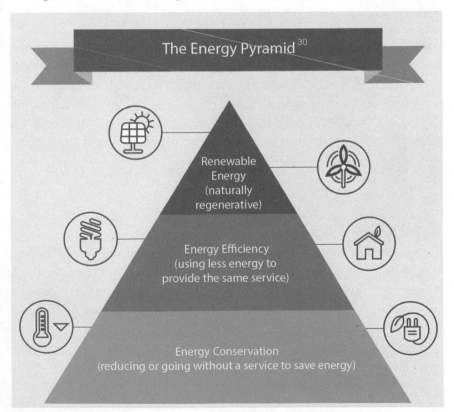

The Energy Pyramid[30]

Renewable Energy (naturally regenerative)

Energy Efficiency (using less energy to provide the same service)

Energy Conservation (reducing or going without a service to save energy)

ExxonMobil is ready. In 2000, the world's biggest corporation supported "access to affordable energy by all and [the] alleviation of poverty in developing countries."[31] Big Oil would love to see government and aid money funnelled to the poor so they can buy oil at prices Big Oil sets. To forestall this, the demand must be for the right to energy from ecologically benign sources at affordable prices.

Ending Fuel Poverty in Britain

Britain partially recognizes energy poverty, calling it "fuel poverty" or the right to a heated home. Britain's regional governments have pledged to wipe out fuel poverty by 2018. Brenda Boardman, at Oxford University, is the lead campaigner, casting the issue around the health concerns of living in cold homes. The ultimate aim is to reduce energy demand.[32] Beyond Britain, the European Union requires all member states to address fuel poverty. Although we live in the second-coldest country in the world, most Canadians haven't thought about or discussed energy poverty issues. Should households have to spend more than 10 percent of their income to keep their homes warm, for instance?

Britain is ahead of Canada on fuel poverty because so much of their housing stock is so bad. Many British houses have no space for insulation. Of those that do, only 35 percent *have* insulation.[33] Largely because of inadequately heated homes, England and Wales saw 24,000 winter deaths in 2011–12.[34] Necessity can spur invention. Each regional government in the United Kingdom has its own fuel poverty plan. Wales has the best principles:

- Work on affordable housing should focus on making running a home affordable.
- Policies on child poverty should be linked to measures to tackle fuel poverty.
- Promoting public health should be linked to ensuring affordable warmth for all.[35]

Grants are available to improve insulation and energy efficiency, including providing heat pumps to some low-income families. The regional governments recognize that incomes must be raised, too. Much better insulation and greatly raised incomes are enough to end fuel poverty.

The Shift to Co-operation and De-globalization

Every successful economic and energy revolution has been accompanied by a cultural revolution that inspires people to change their lives. It's true there must be more windmills and public transit, energy efficiency gains and urban redesign. But even combined, these things won't get us energy security and climate justice. Given the scale of changes needed, we can't do it one light bulb at a time. We need a paradigm shift that challenges the petro-elites' power and their dangerous myths. A lower-energy, inwardly directed shift is the most convincing and most demanding solution. Change our relations with nature and call on different aspects of human possibility, including the empathy gene. Fortunately, such a shift, incorporating but going beyond sufficiency, is underway.

David Korten, a leading thinker on the coming transformation, sees a shift from "Empire" to "earth community."[36] Many citizens' movements subscribe

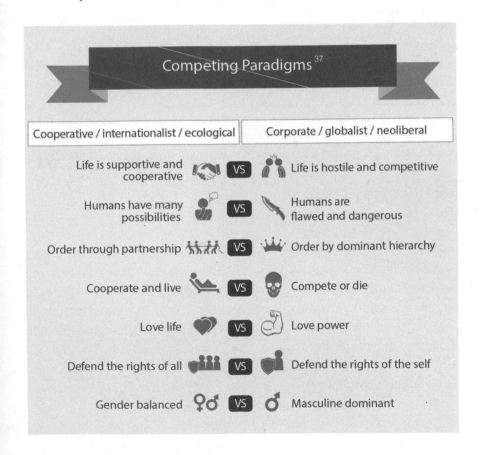

to a version of the co-operative-ecology model. Korten's vision contrasts the corporate, neoliberal model of being hostile and competitive with an existence that is supportive and co-operative.

Korten's earth co-operative paradigm addresses the crises of oil depletion, out-of-control financial and corporate power, and climate change chaos. Three years after the world reached peak conventional oil in 2005, it reached "peak globalization," as Jeremy Rifkin calls it. Many are unaware of that turning point.[38] Unless low oil prices remain low for the long term, it's unlikely the world will return to a hyper-globalized economy. Cheap oil won't be there to underwrite it, nor will the deregulated international financial system that supported global trade and investment before the 2008 financial crisis likely return.

The de-globalizing and renationalizing consequences of the oil and financial crises coincided with the less visible climate change crisis. Real solutions to the latter require cutting carbon fuel use by 80 percent by 2050. Even with growing renewables, intense global trading cannot run on such drastic cuts to carbon fuel use. Economies must renationalize and relocalize: much of the manufacturing that fled North America and Western Europe for the Global South will return. Instead of basing development on exports, even China is increasingly finding its main market at home. The world is returning to more "inwardly directed economies" on ideas that were thrown out when neoliberal globalization took charge. It's time to retrieve and update them to construct sustainable societies built on conservation and greater equality.

Renationalize and Relocalize

After continually denouncing Keynesianism for throttling entrepreneurship and taxing the wealthy, the political and corporate elites who imposed neoliberalism on the world got spooked in 2007–08 when giant banks and investment firms fell like bowling pins. They suddenly reverted to Keynesianism—but only to save themselves. Governments used trillions of taxpayers' dollars to bail out giant banks and corporations deemed "too big to fail." Conservatives George W. Bush and Stephen Harper jointly nationalized General Motors, while ordinary people lost their homes and jobs. It was socialism for the rich and laissez-faire capitalism for the rest. It should have been the reverse: let corporations and banks who advocate cut-throat competition die by their own shibboleths; let the rest create a low-carbon, caring-sharing society based on reciprocal relations that nurture life.

Some of John Maynard Keynes's more obscure but best ideas were those around national self-sufficiency, the perils of borderless capitalism and the failure to pin a value on nature. We need to retrieve them. Keynes's 1933 article on national self-sufficiency goes far beyond the demand-side, counter-cyclical, government-spending ideas he is known for.[39] He begins by explaining that he was brought up "like most Englishmen to respect free trade ... almost as part of a moral law." But Keynes ultimately argues that "most modern processes of mass production can be performed in most countries and climates with almost equal efficiency..... National self-sufficiency... may be becoming a luxury we can afford." National self-sufficiency, Keynes stated, is not an ideal but a means to pursue other goals.

Keynes was critical of foreign ownership because of the divorce between ownership and real responsibility of management. "I am irresponsible towards what I own and those who operate what I own are irresponsible toward me." He was also critical of international financialization. "Ideas, knowledge, science, hospitality and travel—these are the things which should of their nature be international," he argued, but do not internationalize everything, especially not finance: "Let goods be homespun whenever it is reasonably and conveniently possible, and, above all, let finance be primarily national."

Departing from most of his contemporaries, Keynes opposed what we now call globalization. He recognized that political communities need enough autonomy from international forces to gain democratic control over their economies. "We need to be as free as possible of interference from economic changes elsewhere, in order to make our own favourite experiments towards the ideal republic of the future.... A deliberate movement towards greater self-sufficiency and economic isolation will make our task easier." Keynes's critique of liberal economics extended presciently to the environment: "We destroy the beauty of the countryside because the unappropriated splendours of nature have no economic value. We are capable of shutting off the sun and the stars because they do not pay a dividend."[40]

Defend the Local Nationally

It may seem outlandish to suggest that the best way for the most citizens to fight for conservation, energy security and sufficiency is to call on environmental nationalism. Nationalism is often a great people mobilizer but can also be a minefield. Sufficiency advocates often support a "no-growth" economy to limit greenhouse gases and the depletion of non-renewable resources and to preserve nature. These goals are easiest to reach when the population does not grow. But limiting population can be tricky in the Global North, where birth rates are below replacement levels. Under the guise of environmental protection, nativist movements in places like Ireland and Australia promote low population policy when their real agenda is to bar immigrants and minorities.[41]

It's a mistake to steer clear of all types of nationalism to avoid the odious ones. Racist nationalisms are best countered with positive nationalisms committed to inclusivity, anti-racism and support for popular sovereignties everywhere. Nationalism can be progressive and ecological without being anti-immigrant. I've not seen racist "eco-nationalism" in Canada and I hope it never arises. My kind of enviro-nationalism aims to get Canada to abide by international norms on carbon emissions and gain the democratic control and sovereignty necessary to protect residents' energy security, resource inheritance and natural habitats.

Australian philosopher Arran E. Gare promotes a new kind of grand narrative that, instead of one proclaimed truth, recognizes a range of voices, viewpoints and cultures. He contends that while environmental problems are global, it's impossible to orient people for effective action without stories relating to individuals' lives and local problems being integrated into broader eco-narratives: "To be successful, environmentalists will have to... liberate themselves from the destructive imperatives of the world economy. What is required is an environmentalist nationalism which can harness the legitimate anger against global capitalism to carry out the massive transformations necessary to create an environmentally sustainable civilization."[42] Gare's prescription is at least partially borne out. Framing ecological and energy issues is largely specific to each nation and country; mass mobilizations, social organizations and targeted opponents are primarily national and local.[43]

Local action to preserve a loved slice of nature often mobilizes the most people to resist destructive pipeline or oil production projects. The success of "blockadia," as Naomi Klein terms resisters, though, depends on a country's degree of national sovereignty and whether central or regional governments support or oppose their efforts. Does their country have enough sovereignty to defend them, or is it hindered like Canada is by neoliberal agreements like NAFTA? The corporate-initiated and overused concept "think globally, act locally" was developed to neutralize the powerful appeal of resource nationalism. Herman Daly champions national economic sovereignty as crucial to environmental protection: "To globalize the economy by erasure of national economic boundaries through free trade, free capital mobility, and free, or at least uncontrolled migration, is to wound fatally the major unit of community capable of carrying out any policies for the common good."[44]

Canada's Sufficiency Warriors

Attempts by advocates of win-win, ecological modernization to green capitalism have failed. Greenhouse gases continue to rise. More and more species and local habitats are in distress. To transition to a low-carbon future, it's time to return to sufficiency, the Goldilocks idea of enough-and-not-too-much that was dominant among environmentalists in the 1970s. Sufficiency views seem ready to burst back onstage. In his book *When Green Growth Is Not Enough*, Anders Hayden uncovers a wide range of sufficiency advocates in Canada.[45] Hayden was surprised when petro-corporation employees would pull him aside at business conferences to say things like, "There's no reflection on what they've done to this point to damage the planet," and "We have a flawed image of the good life...if everyone is trying to consume like us in the wealthy nations—we don't have three planets."[46]

Many spokespersons for mainstream environmental organizations hold radical sufficiency views but remain silent, fearing biting the hand that funds their jobs and the environmental work they do. Sufficiency gets open play in universities, where professors can challenge the system without endangering their careers. Peter Victor at York University, Bill Rees at the University of British Columbia, and Thomas Homer-Dixon at the University of Waterloo are influential sufficiency voices. Victor uses econometric models to show that Canada cannot possibly reduce greenhouse gas emissions by 80 percent by 2050 if its GDP grows until then. A 3-percent-a-year GDP growth would need a 7.23 percent *annual* fall in intensity of greenhouse gas emissions released from a level of carbon energy burned.[47] That's not possible. Nor is such a pace possible elsewhere unless their economies also stop growing in the GDP sense, where destruction is counted as part of "growth." England's Sustainable Development Commission shows that carbon efficiency gains would have to quicken by *ten times* to stabilize the concentration of carbon at 450 parts per million.[48]

Victor shows that if Canada phases out all high-intensity commodity industries by 2020, its emissions would fall temporarily and then rise again as general economic growth overwhelms environmental gains. Victor proposes *degrowth* as a means to reach "well-being, ecological sustainability and equity." Using standard economists' tools in *Managing without Growth*, Victor shows that Canadians can cut emissions drastically while avoiding mass unemployment, widespread poverty and rising government debt.[49]

Bill Rees is a prominent ecological economist that the *Vancouver Sun* calls one of British Columbia's top public intellectuals. Rees invented the "ecological footprint," a concept used by the European Commission and the World Wildlife Fund to assess humans' demand on ecosystems.[50]

Thomas Homer-Dixon also questions whether growth of material production is desirable or will even be possible. He and Toronto writer Nick Garrison argue that every system is stressed: "In the last century or so, average inputs per hectare of Earth's agricultural land have soared eightyfold....We have converted, literally, petroleum into food and food into billions of people. And we have been able to do this because oil and gas have been ridiculously cheap." But it can't last. "If we're unprepared as we approach empty, the machine won't work anymore, and we'll be left stranded at the side of the road."[51] Growth will end. Homer-Dixon's 2013 *New York Times* article, "The Tar Sands Disaster," caused a big storm in Canada. He argued that President Obama would do Canada a favour if he blocked the Keystone XL pipeline, and that Alberta's tar sands industry undermines Canadian democracy.[52]

Despite the depth of thought from intellectuals like Victor, Rees and Homer-Dixon and diverse support, sufficiency's challenge is too little organizational support to move governments to act. As Rees put it, "The idea of limits to growth is anathema to any government in the world today, which is very much beholden to the corporate sector for its political support."[53] Small organizations creatively highlight sufficiency. Vancouver-based *Adbusters*, a magazine that lampoons the advertising industry's mindless message of *buy, buy, buy,* initiated "Buy Nothing Day," the day after American Thanksgiving. Its call to "Occupy Wall Street" sent protesters off to Zuccotti Park in New York City, and occupations soon spread to many cities and countries.

Several post-carbon organizations modelled on the Post Carbon Institute in California have sprung up in Canadian cities. Concerned about peak oil and climate change, they've worked toward "energy descent." Inspired by France's degrowth tradition, Quebec has a "convivial degrowth" (*la décroissance*) movement that tackles interrelated ecological, social and political crises that also cause a crisis of meaning.[54] Ian Angus, editor of the journal *Climate and Capitalism,* is co-founder of the Eco-socialist International Network. The global crisis, writes Angus, "isn't the result of mistaken policies or ignorance—it is the inevitable result of the way capitalism works." Eco-socialism differs from most earlier types of socialism by fully absorbing ecology premises.[55] Solutions to ecological threats will require the active participation of the great majority of the world's people, eco-socialists insist. Catholic and

Protestant churches also criticize endless expansion, particularly of the Sands. KAIROS, a Canadian ecumenical, faith-based organization, campaigns on sufficiency themes.

Although indigenous peoples in Canada don't usually frame them as such, they nevertheless spearhead some of sufficiency's most effective movements. Clayton Thomas-Muller, a Mathias Colomb Cree, co-directs the Indigenous Tar Sands campaign at Polaris Institute and is a leader in the Idle No More movement. Thomas-Muller's mission is to "protect the sacredness of Mother Earth from toxic contamination and corporate exploration, to support our Peoples to build sustainable local economies rooted in the sacred fire of our traditions." "Sustainable local economies" is another way of saying "*sufficiency*." "In the last thirty years of Canadian environmentalism," Thomas-Muller notes, there has not been a "major environmental victory won without First Nations at the helm asserting their indigenous rights and title."[56]

Despite its diverse advocates and many small networks, sufficiency has, as Peter Victor notes, not yet coalesced into a united movement, found strong institutional support, or made growth a focus of public debate.[57] Sufficiency campaigners have succeeded in curbing particularly egregious practices but not in convincing people to buy less.[58] In Britain, campaigners stopped a third runway at Heathrow, London's main airport; suvs became something of a pariah vehicle; and leading retailers stopped selling gas-powered patio heaters. In Canada, sufficiency advocates have won on much narrower issues—bags, bottles and bulbs—charging a nickel for a plastic bag, banning the sale of bottled water in many public spaces and phasing out incandescent light bulbs.

While symbolically important—and though they have other environmental benefits—each of the victories has cut few carbon emissions. Plastic bag use fell by half in Canada, but Resource Conservation Manitoba calculates that plastic bags in the province use only the equivalent of 1.9 litres of gasoline per person per year. The danger is that "efforts to curb this form of consumption could serve as a symbolic substitute for more significant environmental measures," Anders Hayden argues.[59] Reducing use of plastic bags did not spread the message that sufficiency must replace growth. Paradoxically, Hayden concludes, sufficiency faced "better prospects when it could be linked to increased economic output in some other form....These were instances where the idea of sufficiency, narrowly defined, attached itself to the economic growth model."[60]

If presented as an individual lifestyle choice, consuming less is acceptable, even virtuous. But try to make it government policy and you'll bring down the wrath of the petro-gods, their media allies, and the populist right. As stewards of one-fourteenth of the world's land mass and one two-hundredths of the world's people, Canadians have a huge responsibility to curb greenhouse gas emissions, conserve remaining natural gas and conventional oil supplies, and leave Sands oil in the soil. Powerful vested interests and their ideology block the way. We need the vision and courage to push them aside.

Chapter 9

How Much Is Enough? A Conserver Society

It is entirely reasonable that national governments should have legitimate policies different from those of the oil majors....They are starting to regard their shrinking oil and gas resources as something to be husbanded. King Abdullah of Saudi Arabia recently described his response to new finds: "No. Leave it in the ground....Our children will need it."

—Lord Ron Oxburgh, former chairman of Shell Oil, 2008

While on a trip of recovery and discovery to Australia with my wife, Judith, in 2007, we saw something that smashed our summer idyll. Sydney was hot and very dry. The relentless summer sun eased Judith's joint pains, but harmed the people, animals and plants. A train trip began in beautiful greenish countryside, but that soon changed. For hour after hour we saw abandoned farms, dead trees, even a dead cow in a field. After a decade of drought, the land was dying. As our train pulled into Melbourne we saw a tall, yellow-brick smokestack with graffiti written in large black letters: *There are no profits in a dead world.* Those words have stayed with me. Canadians may think lessons from Australia are irrelevant. We live in a much colder country and always will. But changes in Australia matter. They're harbingers of climate disasters under business-as-usual.[1]

Globally, energy corporations hold 2,795 gigatons of carbon in their proven reserves of coal, oil and natural gas. That's five times the limit climate scientists at London's Carbon Tracker have determined humans can safely emit by 2050. Releasing even 565 gigatons will heat the planet by two degrees Celsius and cause great devastation. But if all the carbon is released, temperatures will rise more and cause utter disaster. "We have to keep 80 percent of the reserves locked underground," argues pioneering ecologist Bill McKibben. Not to worry, advises Jeff Rubin, CIBC's former chief economist in *The End of Growth.* Oil's inevitable price rise will cut its use enough. (Now that's faith in the market's invisible hand.) Rubin fails to ask how much of the carbon energy corporations hold will likely be used. Will it miraculously come under the limits climate scientists set? Most profits don't lie in saving the environment. So the stakes are too high to leave it to the market.[2]

Jeffrey Sachs is a Columbia University economist who promoted capitalist shock therapy in post-communist Eastern Europe but has turned ecologist.

He warns, "The rich will try to use their power to commandeer more land, more water, and more energy for themselves, and many will support violent means to do so, if necessary."[3] An oil CEO who tried to do right for the planet would be fired. Stockholders would demand it. After seeing Climate Tracker's numbers on carbon reserves, Naomi Klein noted that "with the fossil fuel industry, wrecking the planet is their business model."[4]

Proven carbon reserves and prospects for profitably selling them largely determine petro-corporations' share values. That's why BP keeps mucking about in Alberta's Sands, fracks despite horrific side effects, and explores deep oceans—even after its Gulf of Mexico blowout well in 2010. BP branded itself "Beyond Petroleum" in 2001. No longer. It closed its solar division in 2010. The previous year, Shell had shut its solar and wind-power initiatives. As McKibben observes, "The five biggest oil companies have made more than $1 trillion in profits since the millennium—there's simply too much money to be made on oil and gas and coal to go chasing after zephyrs and sunbeams."[5]

Yet in the hope that they'll do that, mainstream environmentalists propose ecological add-on solutions to business as usual. The Alberta-based Pembina Institute does excellent analysis but invariably presents win-win recommendations. The environment will win and companies will profit. The economy will grow while we reduce carbon emissions. It's a feel-good story called "ecological modernization." Ecological modernization recognizes the ecological crisis as a fundamental societal omission, writes Dutch political scientist Maarten A. Hajer. "Yet, unlike the radical environmental movements of the 1970s, it suggests that environmental problems can be solved" by society's main institutions. Environmental management is presented as pollution prevention pays[6] and will not harm economic growth.[7] Ecological modernization never states the obvious. To sustain life on earth, economic growth must slow, dirty projects must close and capitalism must be challenged. The fact that most win-win ecologists are genuinely idealistic makes their alliance with petro-corporations more effective. They impede real solutions by giving false ones credibility.

University of KwaZulu-Natal professor Patrick Bond and his colleagues examined carbon trading in South Africa.[8] Corporation A in the Global North emits carbon and pays corporation B in the Global South to not release what they probably wouldn't have anyway. "The prospect of getting paid real money based on projections of how much of an invisible substance is kept out of the air tends to be something of a scam magnet," writes Naomi Klein. "The carbon market has attracted a truly impressive array of grifters and hustlers."[9]

The problem with ecological modernization is that a tweak here and there and international agreements with voluntary enforcement aren't enough. Not even close. Effective action requires removing the power of petro-elites who profit by blocking climate action. We need a paradigm shift. The ideology of rampant deregulation encouraged a speculative, Ponzi economy that harmed the real economy. That, and oil reaching $147 a barrel in July 2008, caused the Great Recession. The world hasn't yet fully recovered.

Neoliberalism rests on faulty premises that John Kenneth Galbraith characterized as "the rich...not working because they had too little money, the poor because they had much." Neoliberals counsel people to adapt to capitalist economies, the reverse of what economies are for. Economies should be our servants, not our masters. They should enable strong, caring communities, shared wealth, healthy kids, a resilient biosphere, eco-energy security, and democracy as living practice. It took millions of years to turn sunlight stored in dead organisms into carbon fuels. We cavalierly used up the cheapest and easiest part of them in 150 years. If we carry on with business as usual we'll soon suffer a sudden collapse. Or we can move off carbon energy now to land softly.

Much has been written on shifting to a low-carbon economy.[10] The main elements require individual and collective action by people unafraid to challenge the petro-elites' power and their paradigm. Super-insulate all homes and buildings, densify cities and stop urban sprawl. Reorient cities around public transit, walking and cycling, which will also trim waistlines and boost health.[11] Build high-speed electrical trains powered by renewables between big cities. Greatly hike the fuel efficiency of autos and steadily replace them with hybrids and electric cars. Do much more telecommuting, so it replaces much real commuting. Return to low-energy food production, locally grown and consumed as much as possible—a one-hundred-mile diet. "Food sovereignty" brought to our plates by local food producers, not "food security" sold to us by Monsanto.

Various ways to reach a low-carbon future are on offer. Some rely on new technologies like electric cars and renewable energy,[12] others on changing behaviour through pricing nature and high carbon taxes.[13] None challenge business-as-usual capitalism, which assumes the greed of each leads to the greater good of all and that ever-rising consumption helps the economy and brings greater happiness. They fail to address vast inequalities. They'll not likely cut carbon emissions by much any time soon. Most run into the rebound, or the Jevons paradox: efficiency gains often lead to greater energy use. In 1865, William Jevons argued that efficiency gains in coal, the carbon fuel of the first industrial revolution, led ironically to greater overall coal use. Why? When many people conserve a fuel, demand falls and the price tumbles. That stimulates new uses for the fuel, resulting in greater use.[14] The Jevons rebound paradox is at work today. Build more efficient refrigerators and families buy two. Better insulate houses and they supersize them. Efficiency is not enough.

In 2008, 160 environmental organizations from around the world met in Poland to call for "climate justice": "We will not be able to stop climate change if we don't change the neo-liberal and corporate-based economy which stops us from achieving sustainable societies. Corporate globalization must be stopped.... Any 'shared vision' on addressing the climate crisis must start with climate justice."[15] Climate justice recognizes that industrial countries must curb greenhouse gas emissions much more than developing countries. As the Delhi-based Centre for Science and Environment contends, industrial countries "owe their current prosperity to decades of overuse of the common atmospheric space and its limited capacity to absorb GHGS."[16]

Ramp Up Renewables or Cut Energy Use?

Hands-off free marketeers and government-intervening "win-win" environmentalists clash over whether the market will make the transition on its own. Free marketeers see the transition to renewables happening best on its own. For them, the output of non-renewable carbon fuels must eventually fall. The low-hanging carbon fuels are used up first. The cost of bringing on less accessible carbon fuels will rise. This is the "resource pyramid."[17] Rising carbon fuel prices will make renewables more and more competitive so they increasingly replace carbon fuels. Governments should not try to speed things up. They would hurt the process.

But this will take way too long, win-win environmentalists say. They urge governments to make renewables price-competitive now. Tilt the field in favour of renewables through high carbon taxes and subsidized "feed-in tariffs" that pay higher prices for electricity generated by wind and solar than by coal and natural gas, they say. That would make carbon fuels as pricey as renewables, and cut greenhouse gas emissions rapidly enough to allay climate change disaster.[18] Both camps sell the illusion we can simply switch to renewables and continue living energy-intensively indefinitely. They also assume that renewables on a massive scale are benign.

Electricity Generation Efficiencies (%)[19]

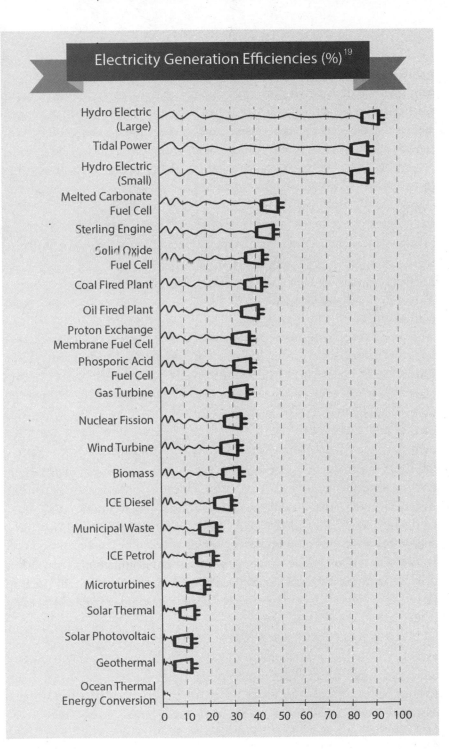

Hydro Electric (Large)	
Tidal Power	
Hydro Electric (Small)	
Melted Carbonate Fuel Cell	
Sterling Engine	
Solid Oxide Fuel Cell	
Coal Fired Plant	
Oil Fired Plant	
Proton Exchange Membrane Fuel Cell	
Phosporic Acid Fuel Cell	
Gas Turbine	
Nuclear Fission	
Wind Turbine	
Biomass	
ICE Diesel	
Municipal Waste	
ICE Petrol	
Microturbines	
Solar Thermal	
Solar Photovoltaic	
Geothermal	
Ocean Thermal Energy Conversion	

0 10 20 30 40 50 60 70 80 90 100

St. Petersburg physicist Anastassia M. Makarieva and her colleagues studied a massive global switch to renewables and concluded that renewables would not stop climate change and could actually harm the atmosphere. "It is commonly assumed that environmental stability can be preserved if one manages to switch to 'clean', pollution-free energy resources, with no change in, or even increasing, the total energy consumption rate.... Climate and environment can only remain stable if anthropogenic pressure on natural ecosystems is diminished, which is unachievable without reducing the global rate of energy consumption." They cite the case of wind power: "When the moisture-laden ocean-to-land winds are... impeded by windmills, this steals moisture from the continent and undermines the water cycle on land. ...Wind power is equivalent to deforestation.... Wind power stations can thus be allowed to exempt less than one percent of total wind power."[20]

The scientists contend that total available renewable energy resources—hydro, wind, tidal and solar—"can in total ensure no more than one tenth of modern energy consumption."

Canadian author Andrew Nikiforuk, the only non-American to win the Rachel Carson Book Award, explains why renewables can't simply replace carbon fuels. "Renewables, by definition, harvest less energy than do densely packed fossil fuels. Like unconventional hydrocarbons such as bitumen and shale gas, they require intensive industrial farming. They also create greater landscape disturbances. To replace a thousand-megawatt coal-fired plant sitting on 1.5 square miles of land with solar panels would require a small city-sized area of 19 square miles. To achieve the same energy gains with wind could take an area three times bigger.... Meeting one-third of the world's current energy needs with wind power would require approximately 13 million towers spaced half a mile apart occupying 3 million square miles: approximately five percent of the world's total land mass... [and] would cost at least $15 trillion."[21] On land, windmills can occupy good farmland or displace forests, the lungs of the world. Wind and sunshine are intermittent. Windmills spin when there is enough wind. Solar panels generate power when the sun happens to shine.

I don't suggest slowing the growth of non-hydro renewables. Far from it. They currently provide only 1 percent of global energy and haven't reached their limits. But let's avoid the delusion that they will permit us to maintain our energy wasteful ways. Canada holds an ace. Dammed hydro power complements wind and solar. Dams can be opened quickly to counter sudden drops in their power. Canada has many dams.

Energy Scarcity: Who Will Have Access?

Jeff Rubin contends that the low oil prices and cheap transportation that underlay globalization in the 1980s and 1990s will never reappear.[22] Even at their lowest levels, oil prices in 2015 are more than double the prices then. When the current oil glut shuts off additions to high-cost production, the world price will resume its upward climb. If it reached $5 a litre, filling your 60-litre tank would cost an astounding $300, deterring middle- and low-income earners from driving to work. But rich partygoers could still zoom off to distant resorts without looking back. The military would try to commandeer enough taxpayer money to fly fuel-guzzling fighter planes 24/7. Who will have access to less energy will be one of this century's great struggles.

Costly oil triggers global recession, and slows or ends economic growth. Since the 1970s, whenever oil has spiked, a recession has hit the Global North.[23] It happened again in the 2008 Great Recession, but it differed in one way. A supply disruption did not cause the shock price of $147 a barrel in July 2008. The world's oil spigots were fully open. Depletion cancelled gains in new oil. Potential oil in the ground is immaterial. What matters is how much can be affordably pumped under politically acceptable limits.

The stratospheric oil prices and collapse of subprime mortgages in the United States triggered the Great Recession. Oil crashed by autumn 2008 to $40 a barrel, then rose to over $100 and fell again. We're in a new pattern. Growth leads to higher demand for oil. The price rises and kills growth until oil prices fall with lower demand. Low oil prices make it uneconomical to bring on most new oil supplies. Most new sources are high-cost. Oil supply shortages eventually spike oil prices again, choking off a short growth spurt, sparking a new recession. Around it goes.

Since 1945, "the global economy has been in recession only 20 percent of the time," Rubin shows.[24] Growth returned after a crash. Now growth is the exception. Growth was capitalism's great promise, enabling it to survive challenges to exploitation, want and depression that brought misery and anger to millions of workers and their families. The rich should not live so well when workers produce the wealth yet go without basics, socialists said. Capitalists retorted that the pie and your slice of it is growing. Forget about who has more. You will be better off in ten years. Your children will be better off than you.

Not anymore. Neoliberal policies gave huge tax breaks to the rich and corporations, and cut public services and social wages, such as unemployment insurance, that lower-income people relied on. In Canada, the richest 1 percent captured almost a third of the income growth from 1987 to 2007.[25] The growing gap sparked Occupy Wall Street protests around the world in favour of the 99 percent. Few on either side of those debates realize how much a growing pie depended on cheap energy. Costly energy scales back living standards, raising costs on most things. It will likely reignite open class conflict. When oil becomes costly again, there will be fewer material goods per person. Social justice issues of who has how much will return with a vengeance. Should the rich few get most of the earth's remaining deliverable carbon fuels and have the right to dump unlimited greenhouse gases into humanity's common biosphere, while many do without necessities?

Tradable National Energy Quotas (NEQS)

I could write a blueprint for carbon energy descent in Canada, specifying how cuts could be made in each sector, but we don't need another top-down academic exercise. What's important is harnessing the ingenuity of the thirty million Canadians over the age of fourteen in figuring out how to conserve energy, protect the environment and cut carbon emissions in ways that best fit the diversity of their communities and lifestyles. Can we devise a workable plan that will mobilize their collective genius? The challenge is to get people to accept the urgent need and see that they can contribute and devise their own solutions as individuals, families and communities.

Most of us don't recognize the enormity of the disasters that climate change chaos and scarce, costly carbon energy pose. Many Canadians harbour frontier myths about limitless resources in this vast, lightly populated land, and guilty hopes that global warming might moderate bitterly cold winters. Nor does faith in new technologies like carbon capture help. The latter won't be able to deliver on a big commercial scale until at least the 2030s. A British parliamentary committee on peak oil wisely concluded that "we cannot solve today's energy problems with tomorrow's new technologies."[26]

Nor can we assume international markets will continue to supply oil at current levels. Oil-exporting countries are burning up more and more of their oil domestically. Texas geoscientist Jeffrey Brown argues that the current top five oil exporters taken as a whole—Saudi Arabia, Russia, Norway, Iran and the United Arab Emirates—will likely provide *no* net oil exports by 2030.[27] They are using or cannibalizing more and more of their own oil. Climate change and the end of cheap oil are not immediate, terrifying threats like Hitler's blitzkrieg. During World War II, Britons and Canadians readily

accepted rationing as a fair way to cope with scarcities: everyone must sac-
rifice for the war effort. But today's wars against climate chaos and resource
depletion have failed to call forth the same "everyone must pitch in" ethic.

Canadians vaguely know about the threats, but their enormity deters
us from taking telling action. Knowledgeable people get little confirmation
from their friends, communities, governments or media. There is a no-crisis
atmosphere. Feeling helpless, most aware people don't alter their lives on the
scale needed—carbon cuts of 80 percent. A few are vegans. Fewer still are
off-grid. Most people will need social and government support to descend the
carbon energy staircase. Most cannot abandon driving until there is plentiful
and efficient public transit, safe and separated bicycle lanes, sidewalks on
suburban streets, opportunities to telecommute, and high-speed, intercity
trains. Individuals can't supply these things.

I advocate a form of personal, tradable national energy quotas (NEQs) that
can guarantee that every Canadian gets sufficient energy at affordable prices
while ensuring that each adult, government and business assumes responsibility for their own emissions. They're very different and much better than
cap-and-trade systems like the European Union's "Emissions Trading Scheme"
where, absurdly, polluting industries are given free allowances to continue
polluting. Unfortunately, Quebec and Ontario have adopted the top-down,
pro-corporate cap-and-trade system. They're unlikely to curb emissions by
much, and certainly not by the 80 percent by 2050 that Canada's Parliament
pledged in 2008. In contrast, with NEQs big polluters must pay.

Tradable energy quotas apply only to carbon energy—oil, natural gas
and coal. Surprisingly, they can also apply to nuclear energy, as we shall
see. They do not apply to renewable energy such as hydro, wind, solar and
geothermal. Those energy sources can and should rise and will partly offset
declines in carbon energy supply. Conceived by David Fleming in Britain,
tradable energy quotas are the fairest, simplest way I've seen to manage carbon
energy descent. They are much fairer and more effective than carbon taxes.
"It is hard to set a rate of tax which changes the behaviour of higher-income
groups," Fleming states, "without causing unacceptable hardship for people
on a lower income."[28]

Tradable quotas are designed for the national level. Critics wonder why,
when all humans share the world's atmosphere, the plan is not global. It
would be "unstable since it would be vulnerable to breakdown anywhere
in the world," Fleming explains, adding that "large-scale problems do not
require large-scale solutions. They require small-scale solutions within a

large-scale framework." We need that large-scale framework to be a tough world agreement to cut carbon emissions and binding targets for each country. Rich countries must cut the most, while the poorest may even be able to raise theirs at first in a plan called "contraction and convergence."[29]

Countries should be forbidden from buying carbon credits in other countries. It's a largely fictitious exercise of fobbing off cuts to your emissions by pretending someone else is doing it for you. We also need an international oil depletion accord like the Rimini Protocol to smooth out the depletion rate and ensure fair access to remaining oil reserves.[30] Some governments hide behind national plans to avoid meeting international greenhouse gas emission targets. But national plans can be the most effective way to reduce carbon energy use: let countries determine the means to fit their unique conditions as long as they're effective. If a binding international accord isn't reached soon, tradable energy quotas will still have to proceed in high-polluting countries like Canada.

Tradable energy quotas require co-operation, democratic input and trust. People have widely sacrificed for each other at national and local levels, but never globally. Nor are they likely to soon. There is no democracy or strong feelings of solidarity above the level of countries. Unfortunately, there is much ignorance and suspicion about distant others. That's why national and local conservation plans are the most effective. At first, Fleming called his plan "domestic tradable quotas" to show their national, non-European-wide application. *Domestic* rather than *national* avoided racial overtones in Britain around who belongs to the nation. But in Canada, *national* has more civic, inclusive and multicultural connotations than in Britain, so I call them "national energy quotas" (NEQS) in Canada's context.

Using acronyms like NEQS avoids the politically toxic term *rationing*. But that is what their opponents will accurately label them. Rationing isn't necessarily bad. An all-party United Kingdom parliamentary committee report on peak oil explained that rationing "contains two intertwined meanings. The first is guaranteed minimum shares for all, the second is limits to what people are allowed to consume. Many of us resent the second, but in times of shortage we cry out for the first."[31]

The parliamentary committee pointed out why rationing through price is a bad idea: "Markets do not distinguish between more and less essential uses of oil—if the global rich are willing and able to pay more to fuel cars and jets than people elsewhere are able to heat their homes or power their hospitals, then the limited supply of oil will flow to the highest bidder.

Demand destruction can be cruel or even fatal for those who can no longer afford energy supplies."[32]

We've rapidly consumed the low-hanging fruit of oil and natural gas. Like it or not, rationing will come. Rationing is not the cause of shortages, but it can be a fair response to them. "It is not a question of *if*," observed Fleming, "but *when* rationing begins, and the sooner we do it, the gentler it will be."[33] We haven't seriously considered rationing since it ended in Canada in 1947. In wartime, rationing reduced consumption in Canada and sent excess food to Britain. Rationing is unpopular today because it's associated with austerity. However, when scarcity hits without a rationing system, people rush to stores to stock up—hoarding. Shelves are soon bare of necessities. Black markets spring up to sell hoarded items at inflated prices. Low-income people quickly suffer. Rationing is meant to fairly distribute scarce necessities. Energy is a necessity, especially in a cold, vast country like Canada.

The 2008 Climate Change Act empowers Britain's government to introduce tradable energy quotas. In was shelved as "ahead of its time" but can be reintroduced without further primary legislation.[34] It got a boost in 2011 when the all-party parliamentary committee on peak oil, including Conservative members of Parliament, unanimously endorsed and enthusiastically promoted them. NEQs address energy insecurity and climate change simultaneously, and in ways that avoid top-down bureaucratic regulation and unwanted surveillance. They allow diverse individual and community solutions. They are intended to spur citizens to "buy into" the plan, building a sense of common purpose by aligning individual and collective aims so actions to reduce carbon and energy usage that benefit the individual also benefit the wider community.[35]

Frances Moore Lappé, author of *Diet for a Small Planet*, wisely counsels that people have to feel they are agents in the plan. "As long as we feel we are cogs in someone else's machine, we can tell ourselves we're not *really* responsible for the impact we do. But when people gain a sense of control over their lives, they're able to acknowledge the implications of their actions and feel good about taking responsibility." In Denmark, she notes, when 175,000 households produced wind power, "producer families accepted their altered landscape. But when government support for distributed production waned and 'larger, purely business investments' came in, the 'public became less willing to look at wind turbines'....What we ourselves choose and create we see through different eyes than if the very same thing had been imposed on us."[36]

How NEQS Work

NEQS are the best way I've seen to encourage people to control how they reduce their carbon energy use. How would NEQS work, would Canadians support them, what are their shortcomings and how can we fix those? NEQS apply only to energy purchases of carbon energy, not to food or other things. Energy purchases cover most carbon emissions because almost everything we buy requires energy to make and get to us. A national carbon "budget" will calculate how much carbon-infused energy Canadians currently consume. It will give separate allotments to individuals, governments and businesses, according to how much carbon fuel each grouping collectively uses. Individuals use about 40 percent in the UK. Individuals in Canada will likely get about 40 percent of Canada's NEQS, too. Governments will get another allotment and businesses another.

Each individual adult gets a *free* weekly carbon-energy allowance (children get them, too), whereas businesses and the public sector must *buy* the remaining units at weekly auctions.[37] All adults, even those without a computer or cell phone, get an NEQ account. The system operates automatically through credit card or debit purchases with no additional action by most consumers. All fuel and electricity carries a "carbon rating": one unit represents 1 kilogram of carbon dioxide or its equivalent in the fuel's production and use. An annual NEQ budget determines the number of units available. It drops each year, like stairs.

Surpluses from units not used can be sold. Conversely, if more is needed, they can be bought. All trading is done at a single national price that rises and falls with national carbon energy demand. Buying and selling would be as easy as topping up a phone card. When individuals buy gasoline or get their electricity bill, the carbon or energy units they use is deducted from their NEQ account. Some consumers won't have any NEQS to offer when buying something—foreign visitors and residents who forgot their NEQ card or sold their quota. They must buy NEQ units in addition to the cost of their fuel purchase.[38] It will be a single transaction.

Canada's carbon energy budget should ideally be set by a committee independent from the federal government to shield it from lobbying by corporations. Governments themselves must buy NEQ allotments, which, as for everyone else, will fall annually. The major role of governments—federal, provincial, local and indigenous—is to support people and businesses in making adjustments in order to live and thrive on declining carbon energy and growing renewable energy. Governments create and oversee infrastructure, city design and regulations like requiring all new houses and buildings to be

Tradable National Energy Quotas (NEQs)

 Each individual (child or adult) gets a free weekly carbon-energy allowance.

Businesses and the public sector must buy the remaining units at weekly auctions.

 All adults will have a NEQ account. Energy credits are deducted automatically through credit card or debit purchases when individuals buy gasoline or get their electricity bill.

Surpluses from energy units not used can be sold and if more units are needed they can be bought. Buying and selling is as easy as topping up a phone card.

 All trading is done at a single national price that rises and falls with national carbon energy demand.

Consumers without NEQs, such as foreigners or residents who have exceeded their quota, must buy NEQ credits in addition to the cost of their fuel purchase.

 Canada's carbon energy budget would be set by a committee independent from the federal government to shield it from lobbying by corporations.

Since the NEQ price will be determined by Canadian demand, it is in everyone's interest to help each other reduce their energy use.

energy-neutral. Governments also set the framework for nurturing growing renewable energy supplies and other measures enabling individuals and businesses to reduce carbon energy use.

Since the NEQ price will be determined by Canadian demand, it is transparently in everyone's interest to help each other reduce individual energy use. The fixed quantity of carbon energy units "makes it obvious that high consumption by one person leaves less for everybody else" wrote Fleming. "Your carbon consumption...becomes my business: I have an incentive to influence your behaviour to our mutual advantage: lower demand means lower prices" of carbon units for everyone.[39] Although the NEQ price will fluctuate, Fleming contended they'll not likely stay high. If demand for units "should rise, then their price will rise; this will tend to encourage people to reduce their demand, and to offer more units for sale." That will tend to reduce the price of units.[40] This may be true in the first few years, but unless businesses, governments and people conserve intensely, the deepening carbon energy scarcity will boost the NEQ price.

NEQS can transform people's outlook and behaviour quickly in ways that top-down schemes can't, Matt Prescott, a British expert on limiting carbon fuel use, contends: "The real value would be to give people a direct personal stake in the problem and its solution, instead of just leaving it to companies and governments. This is empowering and reduces that sense of helplessness. It could bring about radical changes in consumption patterns."[41] The great advantage of NEQS over energy efficiency and energy intensity schemes[42] promoted by Big Oil and the federal Conservative government is that they prevent the Jevons rebound effect, whereby energy intensity and efficiency improve but total greenhouse gas emissions still rise. This won't happen with NEQS unless corporations win exemptions. The energy budget fixes the country's total carbon energy units for the year. The next year the total drops. Businesses will have to closely watch their carbon energy use because they must pay for every NEQ unit. It's adapt or fail.

To provide a stable planning horizon, the carbon energy budget is a twenty-year rolling plan. As year 1 passes, a new year 20 is added. Inflexibility in carbon allowances in the first years is meant to stop businesses and individuals from wriggling out of their carbon energy descent. Some room for adjustment is left in the latter part of the plan.[43] NEQS are normally geared to reducing carbon emissions. But they are equally good at preventing hoarding after earthquakes, wars, terrorist attacks—or when peak oil triggers international oil or natural gas shortages. Canada has no national sharing system in place for such catastrophes, and it's hard to set one up during the chaos of a crisis.

British proponents of tradable energy quotas have thought through the switchover from a reducing-emissions system to an energy-shortage system. Fuels will be "rated in terms of the actual quantity of fuel they represent, rather than ... their carbon content. No transactional procedures would have to change."[44] When the system switches to a fuel budget, a new issue of units will be made for the scarce fuel.

What happens to refuseniks? "If you can't be bothered with the whole thing, you can flog your entire ration as soon as you receive it, but when you buy fuel in future you will automatically be charged for the units necessary to cover your purchase."[45] Refuseniks will pay ever more dearly for their indifference. As the NFQ budget tightens, people "will be encouraged to think up new and co-operative ways of cutting their consumption, such as car-sharing," British investigative journalist David Strahan writes: "The system would soon get people out of their cars altogether, tipping the balance in favour of walking, cycling or public transport for shorter or discretionary journeys."[46]

The late Tony Benn, dean of the British Labour Party's left wing, objected to tradable energy quotas that allow the rich to continue high-carbon consumption and undermine the fair sharing of scarce resources. Benn noted that trading rations was illegal during World War II.[47] George Orwell wrote then that those who got more than their fair share undermined the war effort: "The lady in the Rolls-Royce car is more damaging to morale than a fleet of Goering's bombing planes."[48] We can't assume the sharing ethic of the war years will be there for carbon energy descent. There is not yet a consensus behind it. The gradual nature of the NEQS plan would likely get more support than wartime-like quotas. It will slowly get us to Benn's goal. NEQS would allow the Rolls-Royce lady to drive in its early years, but she would have to stop as the quotas increasingly bit.

Much research has been done about how tradable energy quotas could work, but they haven't been tried anywhere except on tiny Norfolk Island near Australia, population 2,300. We will fully learn their shortcomings only after NEQS begin in a major country. But we've used rationing systems before and can anticipate and pre-fix many likely problems. The NEQ framework is good, but has been conceived as if it will work automatically, without the intervention of politics, influence and power. Will Big Oil simply go away? That's as likely as the polar caps are to suddenly expand again. After getting insights from Adam Ma'anit, a British writer on environmental politics, I pose the following questions about an NEQ plan:[49]

- Will "externalities," international power markets and the full life cycle of products get properly accounted in calculating a fuel's total greenhouse gas emissions?
- Will NEQs be top-down like other proposed solutions? How democratic will the plan be?
- If the plan impacts big-energy users, would it not be subject to lobbying campaigns, alternative proposals, and watered-down measures?
- Is the plan naive about the actual operation of markets?
- Is the regulatory system strong enough to guide the NEQ system?
- Will there be enough of a political consensus?
- Will the plan encourage crime, as other rationing systems have done?
- Will people find ways to avoid paying for additional NEQ units?
- If Canada starts NEQs and the US doesn't, how much fuel smuggling across the border would occur?
- Despite claims about NEQs' redistributive effects, would the rich continue to find ways around them? Would the poor still be priced out of getting the energy they need?
- How susceptible will the system be to electronic counterfeiting and hacking?

These questions are not meant to shoot down NEQs. But tough questions must be asked of any new system. Can we safeguard against the possible shortcomings? It's good to fix potential problems in advance. We don't want the plan discredited before it gets a chance to work. We'd have only one chance to get it right, or else face a future of doing nothing about many Canadians having inadequate access to energy and more frequent and violent climate change events.

Canada has two big advantages. First, we get 59 percent of our electricity from hydro and a growing portion from wind, solar and other renewables. They would not be part of NEQ quotas that limit only carbon energy. Canadians would have to pay for increasingly scarce carbon fuel units and will likely switch to renewable energy. The upside to using less energy includes saving on expenses. Conservation and an increasingly competitive and growing renewables sector should be able to finance themselves without the need for government subsidies because NEQs will tilt the playing field their way.

Supporters of nuclear energy advocate its expansion to fight climate change, claiming nuclear is a non-carbon energy source. Nuclear energy provides 14 percent of Canada's electrical power on top of that from renewables.[50] "As the world's reserves of high-quality uranium ore dwindle," a UK parliamentary report contends, "it has become an open question whether new nuclear power stations would use up more useful energy over their life cycle (in mining, transporting, milling and processing the fuel, building and decommissioning the power stations and managing the waste) than is generated over the power station's lifetime."[51] Let's stick to existing hydro, ramp up other renewables, phase out nuclear and, above all, conserve.

The second advantage we have is that NEQs would make it in Canadians' collective interests to phase out carbon energy exports to free up more carbon energy for domestic use. Producing rather than using carbon energy is Canada's greatest source of greenhouse gas emissions. Declining carbon energy exports will put downward pressure on the NEQ price. Ordinary Canadians would have strong incentives to support the phasing out of carbon energy exports and act as a political counterweight to the influence of Big Oil and workers in the Sands and other carbon-fuel-exporting industries. Canadians' carbon energy use is currently closer to British than to wasteful American levels. Under an NEQ plan, Canadians would benefit from our more constrained use.

Critics will argue that phasing out carbon-energy exports to reduce emissions attributed to Canada is just a calculation sleight of hand and won't cut world greenhouse gas emissions. It doesn't matter where carbon is produced or used. That's true, but if Canada phases out carbon fuel exports and greatly reduces domestic use through NEQs, it will leave much Canadian natural gas, coal and oil in the soil. A major chunk of the world's carbon then couldn't be released. Canadians can't solve the world's climate threats single-handedly, but we can make a difference.

With the end of easy oil, the world economy is beginning to be renationalized and de-globalized. NEQ plans will boost these trends. On balance, that's good. The power and influence of transnational corporations will weaken and provide an opening for bottom-up citizens' democracies built on sufficiency principles to emerge. An NEQ-type plan must be the centrepiece of Canada's answer to the coming eco-energy crises, but it is not a panacea. Additional measures are needed:

- Negotiate with natives as sovereign peoples and respect their stewardship over traditional lands.
- Replace all oil imports with domestic oil to give Eastern Canadians oil security.
- Use only non-fracked, conventional oil to shift Canada to a low-carbon future.
- Get higher economic rents for energy-producing jurisdictions (provinces, territories, indigenous peoples) through agreements with Ottawa. Apply an excess profits tax to fund the shift. Create an agreement between Ottawa and the energy-producing jurisdictions to annually calculate available economic rents on oil and natural gas.
- If any jurisdiction fails to collect 100 percent of excess profits, the federal government will take the remainder, to be used exclusively to fund the shift. That will encourage energy-producing jurisdictions to keep all of the economic rents.
- End natural gas exports to preserve supplies for future Canadian winters.
- End NAFTA's proportionality rule.
- Phase out the Sands.
- Bring in a "Just Transition" program of government support for energy and construction workers in Sands projects, to help them move to other useful work, especially in renewable energy and conservation.

Chapter 10

Solutions: Energy and Ecological Security for Canadians

The major part of the solution…is to reduce oil demand to a level consistent with reasonably anticipated domestic supply.

—Canadian National Energy Program, 1980

[Conservation is] impossible as long as NAFTA continues to encourage, and even dictate, greater production, consumption and pollution.

—Larry Pratt, 2001

Countries will face enviro-energy crises that will force them to transition toward a new balance with nature. How they deal with the crises will affect their people's prospects and security in the 2020s and beyond. We can either control the end of growth or it will happen to us chaotically. There is no third choice. Will Canada continue to export the majority of its finite oil and natural gas supplies, or become a world leader in building a green future? To become the latter, Canada urgently needs an eco-energy plan that must do three things simultaneously: ensure energy security for every Canadian; reduce carbon emissions to protect watersheds, wildlife and people's health; and transition us to a conserver society. The best way to gain energy security is to supply Canadians with their own oil, natural gas and renewable energy. The way to cut the most carbon emissions and protect habitats in Canada and the US is by phasing out Canada's carbon-fuel exports.

The challenge is how to manage the transition. Progressive politicians and mainstream environmentalists fear challenging the consumer lifestyle of working in order to buy things and buying things to be happier. They are even more afraid to confront a capital strike. The following nightmare scenario haunts progressive movements and governments and dissuades them from advocating bold climate change action: if they win office and set up a carbon-lowering plan, profits will fall and capital will pull out; the resulting unemployment could topple an eco-minded government. But history shows that in the right conditions, bold can win.

Changing Direction

Victory over fascism in World War II gave the political left and centre the upper hand in the West for the next thirty years. With the partial exception of the United States, there was a new deal of full employment, public education, universal public health care, and a social wage for citizens in old age, sickness and unemployment. These measures brought much greater equality and security. But the grand bargain struck between workers' movements and corporations also laid the basis for today's carbon-intensive society. The grand bargain saw workers and their allies replace their aim of ending capitalism with annual pay raises, staying in alienating jobs with long work hours in North America,[1] but gaining the middle-class lifestyles of two-car suburban families.

Neoliberalism's ascendancy means elected governments turn over responsibility for public safety to corporations. It's called deregulation. Tragic accidents like the runaway train carrying Bakken shale oil that exploded in Lac-Mégantic, Quebec, in 2013, killing forty-seven people, are the result. Transport Canada allowed the Montreal, Maine and Atlantic Railway (MM&A) to decide how many handbrakes to apply to a stationed train on a descending grade and use only one engineer per train. York University political scientist Leo Panitch writes that MM&A CEO Ed Burkhardt was right when he noted that "the fundamental reason the train wasn't being overseen...was that the 2-percent-higher labour costs entailed in hiring more workers to guard their trains would lead to a company raising their freight rates...[and] losing the business to some other company that wasn't paying for workers to guard their trains." Panitch notes that Burkhardt "spoke nothing less than the truth when he added, 'That's the way it works.'"[2]

After seeing the consequences of neoliberalism and cascading climate change disasters, Canadians may be ready to turn to a bold vision. We can discuss energy security and environmental protection as much as we like, but unless Canadians gain priority access to their own energy, Canada will remain primarily an exporter of carbon fuels to the US and perhaps China. If so, Canada will continue as one of the world's worst per-person greenhouse gas (GHG) emitters. Unless we change course, Eastern Canadians may still depend on oil imports from risky countries when the next international oil shortages hit.

We have seen that world oil output and demand will likely peak soon. Transnational oil corporations know it. Why else would they scrape the bottom of the barrel in their desperate search for more? The American Petroleum

Institute states that if the US drilled everywhere that is now off-limits, it would add only 2 million barrels of oil a day by 2030, about 10 percent of current US consumption.[3] World oil output would rise by only 2 percent. The world will move on. It has to. But don't bet on it buying Sands oil for long.

The Eco-Energy Security Plan

We've examined how Canada has no plans for international oil shortages because Prime Minister Stephen Harper doesn't believe in one. Harper says Iran scares him but he refuses to replace oil imports with Canadian oil to protect Eastern Canadians from Persian Gulf conflicts: let the market provide and protect. By law, corporate CEOs must deliver profits to shareholders. They are not charged with ensuring citizens' well-being. That's the responsibility of governments.

The eco-energy security plan I advocate is straightforward. Use Canadian conventional oil, natural gas liquids and natural gas as transition fuels to get Canada to a conserver society. If we reorient our conventional oil (non-fracked) and natural gas to serve Canadians and end exports, we can fully supply Eastern Canadians and save the rest for high-end uses like making plastics. After transitioning to a low-carbon society, energy security will come from electricity generated almost entirely by renewables.

Canadians' current oil use is 1.7 to 1.8 million barrels per day, while conventional and fracked oil output is about 1.38 million barrels per day. Subtract 360,000 barrels per day of fracked shale/tight oil and the shortfall is about 700,000–800,000 barrels per day.[4] How can we make up the difference? About another 700,000 barrels per day of oil equivalent comes from liquids in natural gas.[5] We don't need Sands or fracked oil. Canada's conventional oil and natural gas output is slowly declining. The eco-energy security plan will reduce oil use at least as fast as they decline. That's not difficult. The hard part is summoning the political will. Canada is not a bit player. With 0.5 percent of the world's people, Canada emits 2.3 percent of the world's GHGs. That's way too much. What Canada does matters.

Natural Gas: The Twenty-Five-Year Rule

We need a Canada-first natural gas plan too. From the 1960s to 1980s, the National Energy Board had a twenty-five-year rule to ensure that Canadians had sufficient supplies for the long run. The board disallowed exports unless suppliers verified there were twenty-five years of *proven* natural gas reserves for Canadians' use. That was *proven*, not *probable* reserves. "Probable" is the slippery language authorities now use. Bring back the twenty-five-year rule that was dropped when the Canada–us Free Trade Agreement was being negotiated. The prevailing deregulation view then, since discredited by the 2008 Great Recession, was that corporations should decide most things, unfettered by governments.

Canada has a decade or so of proven natural gas supplies. They're slowly declining, but Canada won't run out in ten years. Additions are still being found, especially by fracking and in new gas fields in northeastern British Columbia, despite low natural gas prices that inhibit exploration. But new finds are failing to fully replace drawdowns in most years. That's normal for non-renewable resources. Depletion of Canada's natural gas will accelerate if liquefied natural gas (LNG) exports from BC proceed. If Canada brought back the twenty-five-year rule, LNG exports would not be approved.

Canadians are eleventh in per-person use of natural gas in the world. All other heavy users except the Netherlands are net *oil*-exporting countries. Excessive natural gas use accompanies oil surpluses because the countries' governments assume their energy super-abundance will last a long time. Canada shares their deluded cornucopia mentality. About 20 percent of Canada's natural gas is burned to turn Sands bitumen into oil. If it's business-as-usual until 2035, a third of Canada's natural gas output will be used this way, mainly to export bitumen.[6] International agencies attribute the emissions from burning natural gas in the Sands to Canada. The emissions are not considered American even though most of the natural gas is in effect embedded in US imports of the Sands oil they burn. When Canada phases out the Sands, Canada's deemed emissions of GHGs will fall greatly.

A national conservation strategy is needed to lower Canadians' wasteful use of natural gas. Lowering use will stretch out Canada's supplies. In this cold country, we'll need them. Timing is good to revive the twenty-five-year rule on restricting natural gas exports. Shale gas has boosted US output enough that the US will likely become a net exporter of natural gas by 2016.[7] Canada's gas exports to the US are falling. So Canada's reversion to the twenty-five-year rule would have little impact on the US.

The twenty-five-year rule never applied to conventional oil. It should. Canada is running short. We have less than ten years of *proven* supply. Oil is non-renewable and very precious. Three large spoonfuls are equivalent to eight hours of human manual labour.[8] No other energy source on the technological horizon can match its convenience, cost and punch. A twenty-five-year rule would not restrict the export of Sands oil, though. Canada has one to two centuries' supply.[9] The Sands problem is their horrific environmental footprint.

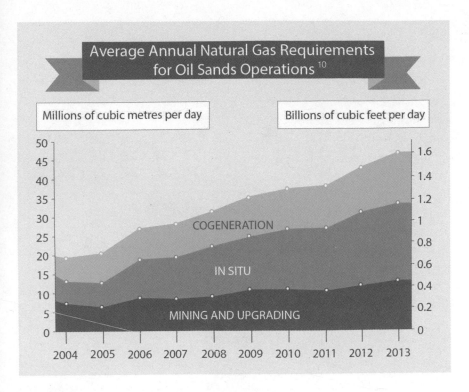

How to Phase Out the Sands

How do Alberta's Sands fit into the eco-energy security plan? Although they give Canada a black eye, they should have the same rules applied to them as are applied to other carbon fuels. If the Sands are singled out, corporations will likely sue Canadian governments for loss of future profits. But if they're phased out by enforcing environmental regulations that apply to all forms of energy, including Quebec's and BC's hydro power, the legal and political case for discriminatory treatment will be much weaker. It will also help lessen cries of regional unfairness.

To continue producing, all fuels should meet low-carbon fuel standards (LCFS) like California's, as well as tough environmental regulations and enforcement to protect local habitats, wildlife, watersheds and human health. They should be allowed to carry on only if they continue to meet the incremental targets. Low-carbon fuel standards require oil companies to reduce the average amount of their carbon emissions across their fuel supplies. In California, oil companies have to sell increasingly cleaner fuels to meet the rising standard. California's LCFS apply at the retail level. Canada should apply them to the production side, much like corporate average fuel efficiency

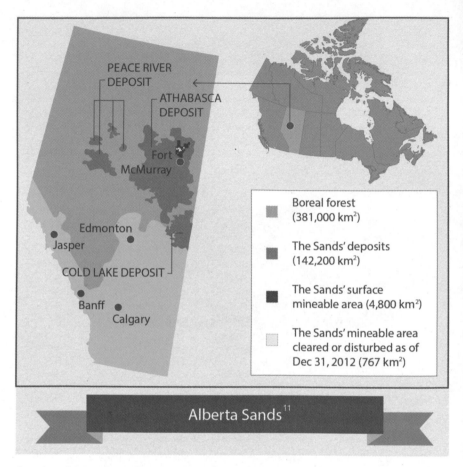

PEACE RIVER DEPOSIT

ATHABASCA DEPOSIT

Fort McMurray

Edmonton

Jasper

COLD LAKE DEPOSIT

Banff

Calgary

Boreal forest
(381,000 km²)

The Sands' deposits
(142,200 km²)

The Sands' surface
mineable area (4,800 km²)

The Sands' mineable area
cleared or disturbed as of
Dec 31, 2012 (767 km²)

Alberta Sands[11]

(CAFE) rules apply to auto companies. Oil corporations could be assessed on their average emissions of GHGs for oil they produce in Canada.

It's unlikely Sands producers will be able to lower carbon emissions to conventional oil levels, because bitumen must be heated to separate the small portion of oil from the mounds of sand. If natural gas heats the bitumen, the consumed gas emits many more GHGs than in conventional oil extraction. Producing oil, not consuming it, is Canada's biggest source of GHGs. Nuclear power can heat bitumen, but that wouldn't necessarily lower GHG emissions if the life cycle of extracting uranium and decommissioning the nuclear plants at the end of their lives is accounted for. As well, Sands operators would have the formidable challenge of showing they could safely dispose of hundreds of years of radioactive wastes.

It is almost impossible to stop Sands production from damaging northern Alberta and other habitats that Sands oil transverses by pipeline. If the

Sands remain more damaging than conventional oil, their production must be capped and then phased out. Projects that have paid their capital costs should be closed first. Once all Sands projects are shuttered, Canada will be much closer to meeting international GHG targets. Simultaneously, Canadians need to steadily reduce carbon energy use.

Interprovincial, All-Canadian Electricity Grids

Electricity will be the energy transmitter of the future. To effectively combat climate change, electrical power must be generated almost entirely by renewables. To get there, we need to shift from exporting energy to producing interprovincial and local power. Canada exports only 10 percent of its electricity generation—all to the US.[12] Although minor, these exports are a big obstacle to getting to the new paradigm. If a province exports *any* electricity to the US, its whole operation in Canada is subject to regulation by the US Federal Energy Regulatory Commission (FERC). Economist Marjorie Cohen calls it "US regulatory imperialism."[13] FERC aims to end public control in the US by privatizing government-owned utilities, and to deregulate electricity by separating power generation from transmission. Since 2000, FERC has called for the creation of large transmission areas or Regional Transmission Organizations (RTOS). Each RTO acts like a private company with full authority. Power utilities have no say over prices, system reliability or who gets the electricity.

When a province exports *any* power to the US it is subject to US-controlled RTOS and FERC. The US, then, not Canada, in effect regulates provinces' electricity. FERC demands a continental power system and will resist the better ecological and democratic alternative—interprovincial, all-Canadian e-grids. The way to break FERC's extraterritorial power is for every province to stop exporting *any* electricity to the US. This would hardly affect the US, as it gets a tiny portion of its electricity from Canada. The problem would be resistance from the provinces. Provincial power companies get better prices by selling electricity to the US than to sister provinces. Although six provinces lack sufficient hydro sources and four have huge surpluses, there are insufficient interprovincial grids to transfer power between them. There are only bilateral agreements between some provinces. If Canadians are to use less carbon energy, we need to share hydro power.

With half a percent of the world's people, Canadians consume more hydroelectric power than any other country.[14] Thirty-five million Canadians use 21 percent more hydroelectricity than 314 million Americans, almost eleven times as much per person. We get 60 percent of our electricity from hydro, compared to their measly 6 percent.[15] Canada is fortunate to be able to base its transition to a low-carbon future on prodigious hydropower. Canada cannot rest on its laurels, though. We get 17 percent of our *total* energy from hydro. Norway gets 44 percent.[16] Most of Canada's energy still comes from oil, natural gas and coal. Additions to hydro power should come only from run-of-the-river turbines and tidal turbines, not big dams. Efficiency gains and conservation can stretch hydro power much further.

Lacking interprovincial e-grids to share hydro power among provinces is disastrous for GHG emissions. Provinces with little hydro still generate their own electricity, largely from carbon-rich coal and methane-laden natural gas. Although it does not stay in the air as long, methane is an even more potent GHG than CO_2.[17] It makes sense for hydro-have provinces to share power with hydro-have-not provinces. Alberta and Saskatchewan are the worst offenders, emitting four times the per-person GHGs of other provinces. With 15 percent of Canada's people, the two provinces produce 45 percent of Canada's GHGs.[18] Much of the excess comes from oil extraction disproportionately located there. But power generation is a major culprit, too. Coal and natural gas generate almost all of Alberta's electricity and three-quarters of Saskatchewan's. Ontario has five nuclear generating plants and New Brunswick one. Ontario, Nova Scotia and New Brunswick produce a major portion of their electricity with natural gas, while the latter two also use coal.[19]

It needn't be this way. Alberta could get surplus hydro power from BC if the latter ended exports. Manitoba could supply Saskatchewan and northwestern Ontario. Ontario could get lots of hydro from Quebec, and Newfoundland can supply the Maritimes. The hydro-rich provinces—BC, Manitoba, Quebec and Newfoundland and Labrador—have enough surplus hydro to top up the other provinces' power supplies. Interprovincial e-grids could use high-voltage direct current to reduce energy losses over long distances.[20] As more people adopt electric and hybrid vehicles and use subways, light rail transit and inter-city high-speed rail powered by electricity, demand for electricity will shoot up. The growing load can be met by adding in non-invasive hydro, wind, solar, geothermal, biomass and other renewables to interprovincial e-grids. At the same time, we must lower power demand through district heating, better-insulated buildings and switching from cars, even electric cars, to public transit, walking and cycling.

Tesla's low-cost, compact home batteries may so decentralize the production of power that regional e-grids will eventually become redundant. Homes and buildings could supply their own solar power and store it for times when the sun isn't shining.[21] That would be a profound revolution, but is likely to be decades in the making. In Canada, Tesla's battery may get its main use initially as a backup to power homes during system outages. Over time, home and building batteries and solar could revolutionize the delivery of power, especially as electric cars catch on.

Oil is both an ally of autos and their adversary. Everyone knows that most cars can't run without oil, but few know it's difficult to build cars when a petro-economy is riding high. It's hard to make other things as well. The loonie rises too high. Will making things return to Canada as the Sands boom cools? Or have Canadians regressed to their colonial role as hewers of coal and haulers of oil? Daniel Trefler, a University of Toronto business professor, argues that Canada faces a trade-off. Do we want to be an innovation-based economy or a resource-based economy? Trefler called the decimation of Canada's innovation-based economy, centred in manufacturing, the "loonacy" of a high dollar.[22] "As the commodity-driven loonie rises," argued Trefler, "it becomes too expensive to produce innovative goods and services in Canada.... [Companies] exporting such world-beating products as asset management services (Manulife) [and] regional jets (Bombardier)... have been forced by the strong dollar to move at least part of their operations to lower-cost jurisdictions. That's Dutch Disease. That's the weakness of a strong loonie."[23]

Pushing the Plan

The security plan will ignite battles. Some may look like replays of the regional battles over the 1980 National Energy Program (NEP). Others won't. A major difference is that Alberta is no longer on the outside "wanting in." Prime Minister Harper hails from Calgary and portrays Big Oil's vested interest in unlimited exports as being in Canada's national interest. Since the Pierre Trudeau NEP era, political power has shifted from Quebec to Alberta. Stephen Harper runs only the second majority government in Canadian history that does not depend on Quebec seats. The first, Robert Borden's, was elected in English Canada during the bitter split with Quebec over conscription in World War I. No special circumstances led to Harper's English Canada–only victory in 2011. A Quebec versus Alberta fight can easily recur, though.

In 2007, Peter Lougheed predicted a looming war between Alberta and the federal government over pollution caused by Sands development that will far surpass previous conflicts over the NEP and repatriation of the Constitution.[24] He also foresaw considerable battles over these issues within Alberta,[25] which was later borne out by Notley's "Orange Wave". It follows that the stronger the support within Alberta from citizens' groups and from the provincial government for an eco-energy security plan, the less its depiction as anti-Alberta will stick. Governments and citizens have to lead.

Although opponents will brand it NEP 2, a new eco-energy plan will be different. While the NEP promoted rising oil production, an eco-energy plan focuses on lower production and use based on the principle that a unit of carbon energy saved is better than one discovered, produced in damaging ways and emitted. The upside is that saving a unit of energy is much more labour-intensive than producing one. That will mean more jobs and they will be spread much more evenly across Canada. All regions need retrofitted buildings, for example. It is much better than the carbon economy that often overheats Alberta and leaves large pockets of jobless Canadians elsewhere.

Pierre Trudeau's NEP encouraged the expansion of Alberta's Sands and new conventional oil in the "Canada lands"—areas under federal control, including the Yukon, Northwest Territories, what is now Nunavut and offshore. These sites did and still do carry great environmental risk. In contrast, the eco-energy plan would not divert oil production away from Alberta, but would rely on conventional oil and natural gas liquids from Alberta, Saskatchewan, BC and Newfoundland. The eco-energy plan would displace all oil imports by turning the proposed Energy East oil line into a line that carries non-fracked, conventional oil for use throughout Quebec. There would be no new pipeline to New Brunswick. All oil imports to Atlantic Canadians would be replaced by Newfoundland's conventional oil. This option was unavailable at the time of the NEP because Newfoundland oil had not yet come on-stream.[26]

In contrast to the NEP, the eco-energy plan would keep the international oil price. This means no complex rules to maintain a low Canadian oil price and no oil export tax that so enraged the Alberta and US governments. Keeping Canada at the high world oil price would promote conservation and leave more room for provinces, territories, First Nations and Ottawa to collect much higher royalties to fund the transition. Trudeau's NEP promoted private Canadian ownership through tax breaks and subsidies that favoured Canadian capitalists over foreign ones, which angered Washington and medium-sized US oil corporations with operations in Canada. In the eco-energy plan, all energy companies would become public interest, not-for-profit entities. Incorporation rules would require owners to have location commitment to Canada, invest only here, and be owned only by Canadian citizens and permanent residents.

In the eco-energy plan there must be no assertion of federal dominance over the economy, nor promotion of the governing party. Those features of the NEP eroded significant support. The NEP predated scientists' warnings about GHGs and climate change disasters. In contrast, the eco-energy plan makes curbing GHGs central. Progressively tougher environmental standards and tradable national energy quotas would ensure that Canada meets Parliament's 2008 target of an 80 percent drop in GHG emissions from 1990 levels by 2050. The NEP did not face NAFTA's proportionality rule. The eco-energy plan does, so proportionality must be removed.

At the time of writing, the regional clash predicted by Peter Lougheed hadn't occurred because Harper's government has not protected the environment. Bill C-45, passed in 2012, gutted many environmental safeguards and sparked the Idle No More movement. Big Oil has had no cause to complain. Nor are they likely to before a change of government in Ottawa. The coming regional clash has been postponed. When Ottawa seriously addresses climate change and Canadians' energy insecurity, the conflict will likely come.

Alberta's Reaction to the Necessary Change

Don't expect a cool, rational debate. Although the battle will be mainly a class conflict between Big Oil and Canada's 99 percent, the media may not portray it that way. Big Oil and its supporters, especially in the Conservative Party, are bound to charge that the eco-energy security plan directly attacks Albertans and threatens Confederation. Canada is so decentralized that when a resource dominates a province like petroleum does in Alberta, resource transnationals try to capture or cajole the provincial government into defending their interests. If captured or pressured, the transnationals then advocate provincial rights to thwart effective federal action on energy and the environment.

Alberta's historical grievance of not being heard by Ottawa was true but isn't anymore. Alberta is Canada's richest province, spawns prime ministers and determines that Canada's national economic strategy revolves around Sands exports. How more "in" can Alberta get? However, when the federal Conservatives lose power, expect to hear the Western alienation refrain again, especially if Ottawa's next government gets serious about Canadian environmental and energy security. The petro-elites will also likely pull out the provincial-control-over-resources argument to foil any national plan to curb GHGs, encourage conservation and bring energy security to every Canadian. They will call it restoring provincial control over areas Ottawa has illegitimately encroached on. A look at past tactics is a good guide.

They could pursue a version of the firewall letter issued by Stephen Harper and five other prominent right-wing advocates in response to the Chrétien Liberals' perceived unfair attacks on Alberta in the 2001 federal election: "[If] an economic slowdown, and perhaps even recession, threatens North America, the government in Ottawa will be tempted to take advantage of Alberta's prosperity, to redistribute income from Alberta to residents of other provinces in order to keep itself in power. It is imperative to take the initiative, to build firewalls around Alberta, to limit the extent to which an aggressive and hostile federal government can encroach upon legitimate provincial jurisdiction."[27] The firewall authors proposed that Alberta withdraw from the Canada Pension Plan, create an Alberta provincial police force to replace the RCMP, and collect its own personal income taxes.

Those with the money and power will also repeat the following arguments ad nauseam against an eco-energy plan. First, if the Sands are still expanding, it will be contended that they are Canada's economic engine and major job creator with spread-effects to other provinces and that they attract jobless Canadians from elsewhere to work in Alberta. Second, it will be argued that Alberta's energy industry sends billions of tax dollars to Ottawa with which it finances equalization payments to poorer provinces: Alberta's Sands are good for national unity and Canadians' well-being everywhere. Third, petro-interests will contend that too-high carbon taxes or cap-and-trade prices will punish Alberta by recycling dollars made in Alberta to other parts of Canada.[28]

Heated rhetoric will likely accompany firewall-like strategies. Saskatchewan Premier Brad Wall's response to Thomas Mulcair over Dutch Disease is likely a rehearsal of what we'd hear if Ottawa were to adopt a green energy security plan. Addressing 2,200 supporters in Regina in 2012, Wall charged Mulcair with "espousing wealth transfer policies dressed up as environmental policies that would kill jobs in Saskatchewan, that would drive energy rates in the province, that would threaten our opportunity and our ability to contribute to this country like we've never contributed before." He said Mulcair's "problem is with this premier and it is with this government and it is with the province of Saskatchewan."[29]

A federal government with the courage to bring in a good eco-energy plan must expect such virulent arguments. It would help if citizens' movements that are used to being critics quickly come about to help the government ward off such attacks. Most Albertans and Saskatchewanians are as strongly attached to Canada as to their provinces. They should be wary of appeals to turn their justifiable regional pride into supporting narrow petro-interests—

and the associated environmental devastation they will have to live with long after Big Oil departs. Progressive Albertans and perhaps the provincial NDP government can be especially influential in countering the charge that being pro-Canadian eco-energy security is anti-Albertan. Even some in Alberta's oil patch get that the Sands won't last forever: "Thirty years out, we won't be burning hydrocarbons the way we do today," stated Clive Mather, former CEO of Shell Canada. "Our enemies may not be at the door yet, but they are beginning to circle around Alberta."[30]

US Responses to Phasing Out Canadian Oil Exports

If Canada doesn't end oil exports, there won't be enough domestic conventional oil for Canadians. When I propose to Canadian audiences that Canada end oil exports to the US, fear is the common response. People say the US wouldn't allow it, or even that they would invade. I reply that we need to be boldly ready to stand our ground, as the Americans always do where their interests are concerned. We can gauge America's likely reaction by assessing recent statements by US officials about relying on Canadian oil, and looking at US responses when Canada cut oil exports in the 1970s. After the international oil supply crisis in 1973–74, Canada quickly built an oil pipeline from Sarnia to Montreal—today's Enbridge Line 9—to replace oil imports to Montreal; it carried Western Canadian oil that had previously been exported to the US.

It is remarkable that in an oil supply crisis during high Cold War tensions, the US reacted mildly to a decision by its largest foreign oil supplier, Canada, to end oil exports. Washington's restraint contrasted with its aggressive response several years later to the Canadianizing of the oil industry under the National Energy Program. The lesson may be that Washington values the freedom of US corporations to invest and profit abroad more than in getting Canadian oil. Would the US likely act similarly today? A thoughtful paper by political science graduate student Tanya Whyte gives some answers.[31] Although building what is now Enbridge Line 9 meant reduced Canadian oil exports to the US, she depicts Washington's reaction as "negative but relatively mild."

The US had been the world's great oil-producing and -exporting country but became a net oil importer in 1970.[32] When the oil crisis hit three years later, Canada was America's largest foreign oil source. At 21 percent of US imports, they were almost half as high as imports from OPEC countries. Canada did not regain that proportion of US oil imports until 2010. The US was as dependent on oil imports from many countries then as it is now—about

35 percent. While Canada was phasing out oil exports, US oil imports doubled between 1971 and 1977.[33] Why was US criticism so muted? Three reasons have been offered. First, the US government had also pursued aggressive energy security policies to mitigate the 1973 oil supply crisis, making it hard to criticize Canada for doing something similar. Second, Washington was obsessed by Watergate. Third, Washington worried that a strong US reaction could provoke Canada to become even more energy-independent.

A year before the 1973 Arab oil boycott, President Richard Nixon told Canada's Parliament that Canada should be independent. The US wanted to disengage from the burdens of empire after the Vietnam War sapped America's economy and spirit. "It is time for us to recognize that we have very separate identities; that we have significant differences; and that nobody's interests are furthered when these realities are obscured....Our policy towards Canada reflects the new approach....The Nixon Doctrine...rests on the premise that mature partners must have autonomous independent policies; each nation must define the nature of its own interests [and]...its own security."[34]

Nixon's "Project Independence" promoted US energy self-sufficiency. Carter was even more assertive about it, focusing on conservation, cutting oil imports, and boosting taxes on oil corporations. Vice President Walter Mondale visited Canada to discuss energy co-operation with Trudeau in 1978. *Washington Post* editor Ben Bradlee said, "[I] don't think most people down here have changed their perceptions of Canada....You've raised your prices for gas and oil but then so has everybody else."[35] Was the Watergate scandal and impending impeachment a reason for America's mild reaction? Unlikely. Nixon resigned in August 1974, before Canada announced it would end oil exports by 1982.

The third explanation for Washington's mild reaction is that it worried that pressure would provoke Canada to assert even more energy independence. The US deputy secretary of the interior doubted the US could get greater Canadian oil exports: "They have some serious internal problems; they have a [Liberal] government that's in power only by a sort of ad hoc arrangement. It's the [NDP] tail wagging the dog."[36] The NDP advocated a Canadian-owned oil industry, much of it by public ownership. Pressing Ottawa would spur the NDP and force an election featuring anti-Americanism. It was better to refrain from comment and leave opposition to the Sarnia-to-Montreal oil pipeline to Alberta, the US official advised.

Ottawa held Canada's oil price below the rapidly rising international one. It lowered the price of imported oil in the East by using revenues it gained from a tax on Canadian oil exported to the US. Canada's oil export tax upset Washington more than falling Canadian oil exports. Washington threatened to block overseas oil that reached Montreal by pipeline from Portland, Maine. The threat was real until the oil pipeline reached Montreal in 1976. The threat showed Canadians the danger of relying on oil piped through the US. The US could have played a bigger card: cutting off Western Canadian oil that reached Ontario via pipelines through Great Lakes states. It didn't, because Canada convinced Washington that diverting Western Canadian oil to Quebec was the reason for the oil export tax.

The US Energy Information Administration issued a calm report showing the US could easily cope.[37] Private industry would compensate for reduced oil imports from Canada by building new pipelines from Texas to the East Coast, or unloading Alaska oil. The volume of oil exports Canada planned to end—800,000 barrels per day—was much less than today's 2.3 million barrels per day. But US oil production was falling then, whereas today it's rising. Canadians' knee-jerk fear—that the US would invade or declare economic war if Canada phases out oil exports—is likely overblown.

After all, Mexico nationalized the largely US-owned oil industry in 1938 and got an exemption from NAFTA's proportionality rule fifty-five years later. No invasion. In 2002, the US helped plot a failed coup in Venezuela, the country with the world's largest oil reserves. The US is hostile to Venezuela's socialist government but trades with it anyway. PDVSA, Venezuela's national oil company, owns Citgo, a major refiner and marketer of gasoline in the US. The US also trades with China, whose government it does not like. Just because Washington doesn't like a foreign government's policies doesn't mean it will aggressively retaliate.

Much will depend on the stance of US authorities and on whether the president is Republican or Democrat at the time Canada takes bold action on energy security and climate change. Most Republicans favour boosting domestic and international oil production, and oppose strong regulatory and tax measures to cut Americans' super-wasteful use of carbon fuels. American Democrats are more likely to be amenable. They tend to take climate change seriously and lean toward curbing domestic demand to end US dependence on Middle East oil. But it's not as simple as Republican president bad, Democratic president good. It was Republican president Nixon who issued Canada's declaration of independence in Ottawa. Democratic President Bill Clinton refused Prime Minister Jean Chrétien's request to rescind NAFTA's proportionality rule after Chrétien's 1993 election victory.

It's not just a matter of who occupies the Oval Office when Canada declares sustainable energy independence. Changes in attitude toward Canadian energy occur during a president's regime. When Vice President Dick Cheney released the US National Energy Policy in 2001, Canada was not listed among the eight countries whose oil reserves the US counted on most for imports.[38] US energy security rests on getting oil from diverse sources. Four months before 9/11, George W. Bush declared that "over-dependence on any one source of energy, especially a foreign source, leaves us vulnerable to price shocks, supply interruptions, and, in the worst case, blackmail."[39]

By 2003, though, Alberta's Sands were internationally recognized as the world's second-largest recoverable oil reserves. Washington took notice. By 2005, the US fully counted on the Sands as a major way to end US dependence on Middle East oil. These changes occurred during George W. Bush's watch. Samuel Bodman, Bush's energy secretary, flew over the Sands. "No single thing can do more to help us reach the goal of reducing US dependence on unstable regions ... than realizing the potential of the oil sands in Alberta," he crowed.[40] The day after Stephen Harper won his first minority government in 2006, he received advice from the "Oil Sands experts' group workshop" in Houston, involving officials from all three North American countries: "The Canadian oil sands ... will be a significant contributor to energy supply for the continent."[41] The oil experts in Houston deemed Sands oil a continental, not an Albertan or Canadian, resource. Here were Texas oilmen and Washington advising Canada what to do with Sands oil.

The US Energy Independence and Security Act and the US Energy Policy Act assume Canadian oil is American. The latter says the "Secretary of Defense shall develop a strategy to use fuel produced in whole or in part from coal, oil shale, and tar sands ... and refined or otherwise processed in the United States in order to assist in meeting the fuel requirements of the Dept. of Defense when the Secretary determines that it is in the national interest."[42] The wording is unclear. When it mandates that the oil meet US military needs, does it refer to undeveloped tar sands in Colorado, Wyoming and Utah, or to Alberta's Sands? The Energy Policy Act recommends "initiating a partnership with the Province of Alberta, Canada, for purposes of sharing information relating to the development of oil from tar sands." The wording is highly unusual, treating Alberta almost as if it is a US state. International protocol dictates that countries deal with each other's central government only, never directly with a foreign country's sub-units.

A particularly outrageous section calls for a plan to make "North America" energy self-sufficient by 2025. The act mandates a US "Commission on North American Freedom" with a directive for the chair to select staff from qualified citizens of Canada, Mexico and the US. The commission is to report to Congress on "recommendations regarding North American energy freedom." US legislators appointed themselves as the sole body to deal with the "continent's" energy. When did Canadian and Mexican resources become American? Some call Obama "Bush lite." That may be an understatement. Bush never started the Commission on North American Freedom. Obama did in 2010.

When former Alberta premier Peter Lougheed met US Treasury Secretary John Snow in 2005, Lougheed said Canada should consider exporting oil to China because "it's Alberta's oil." Snow was surprised and upset. NAFTA forbids sales of Canada's surplus oil to China, and Canada is obliged to sell its oil to the US, Snow asserted. Lougheed rejected this and insisted that Alberta can resist US pressure for guaranteed oil: "Sure. We own it."[43] Lougheed worried that the US was getting too possessive about Alberta's Sands. "We should be more straightforward with the Americans that our development of the oil sands will be to our benefit.... I keep reading in the American media that somehow their whole energy problems are going to be solved by the Canadian-American oil sands and that's not a good thing."[44]

The Canadian Association of Petroleum Producers (CAPP) warns that if the US doesn't buy Sands oil, China will. The US would be more dependent on Middle East oil but the result would be the same for the climate.[45] The argument makes sense until you realize that it would take years to get Sands oil to Asia in volumes large enough to rival Canada's oil exports to the US. It would take all the pipelines in combination: Northern Gateway, twinned Kinder Morgan, Enbridge Line 9 and TransCanada's Energy East line to equal the current capacity of oil pipelines from Canada to the US. Before much Sands oil can flow to the Pacific and Atlantic, and before they are approved, built or repurposed, the lines must first overcome great citizen resistance.[46] In the short run, if more Sands oil can't get into the US, the Sands can't expand.[47]

Is Obama's government as focused on Alberta's Sands for gaining US energy independence as Bush's was? Obama's first energy secretary, Steven Chu, a strong advocate of shifting off carbon fuels, was ambivalent about the Sands. "Certainly Canada is a close and trusted neighbor and the oil from Canada has all sorts of good things. But there is this environmental concern," Chu said. Ernest Moniz, Chu's successor, is called a "frackademic" by his detractors and is a supporter of Big Oil.[48] When she was secretary of state, Hillary Clinton said, "We're either going to be dependent on dirty oil from the Gulf or dirty oil from Canada....Until we can get our act together as a country and figure out that clean, renewable energy is in both our economic interests and the interests of our planet, [the US will remain dependent on oil]."[49] A dramatic picture in *National Geographic* magazine in 2009 was many Americans' first impression of the Sands. Since then, pipeline spills of Alberta bitumen in the US have made imports of Sands oil an American issue.

What are the differences between Canada's oil self-sufficiency policy in the 1970s and now? Jimmy Carter was more determined to gain US oil independence through conservation than Obama is. This is a puzzle, because peak oil and climate change have only recently been widely seen as serious threats. The September 11 attacks had also not yet occurred in Carter's time. The attacks may have heightened US concerns over energy security, but they have not prodded the US to take strong enough measures to cut oil use down to domestic production levels.

Stephen Harper talks about Canada as an energy superpower and the importance of Canadian oil for the US. But superpowers don't rely on only one foreign oil source. The US energy security strategy is to spread risk. When he was a senator, Obama said it doesn't matter if the US imports oil from "budding democracies [or] despotic regimes....They get our money because we need their oil....Every single hour we spend $18 million on foreign oil."[50] Whether it came from the sands of Iraq or Alberta, Obama's point was that dependence on foreign oil is *bad*. Richard Haass, president of the influential Council on Foreign Relations, wrote that Alberta's oil sands are not critical to US energy security.

These views among influential Americans indicate that Washington may take phasing out the Sands and Canadian oil exports in stride regardless of the political party in the Oval Office. The US response to reductions in Canadian oil exports was subdued in the 1970s because the US was doing something similar at the time. It's crucial to frame Canadian eco-energy policy around "security." I like the sound of a Canadian "Energy Independence and Security Act," simply substituting *Canadian* for *American*. It will be difficult for Washington to declare that a policy used by those running an empire is different when it's established in an energy satellite like Canada.

Chart a Canadian Course

This is a compelling plan to ready us for the time when oil and natural gas become scarcer and pricier again, as they will soon. It's also a way to stop destroying so many of Canada's habitats and greatly lower its greenhouse gas emissions. Canadians are among the world's most highly educated people, fully capable of doing higher-order things than digging up carbon fuels and shipping them out raw. This is the moment to shift. It is the best time to phase out natural gas exports to the US. The US is becoming self-sufficient in natural gas. We need Canadian natural gas for the long run; the US doesn't. While the US expects to continue oil imports, they are falling as US production rises and demand falls. Corporate average fuel efficiency standards are driving down US oil consumption. In the next few years, the international oil price will likely rise and also curb US oil demand. Canada should seize this moment to start phasing out its carbon energy exports and chart a Canadian course.

The world's eco-energy problems are daunting. But with steely determination we can overcome them. Challenges are not what deter humans. We are built with the brains and ingenuity to overcome. But we can get sidetracked by bogus ideas, or get discouraged by prophets of doom. Yes, the petro-elites have enormous power and the mainstream media is mainly controlled by those with big money allied with Big Oil. So are most governments. But we have one thing on our side: this is a struggle for our lives, those of our children and of all other living things. If we can win over most people to the cause, we will have the numbers.

Frances Moore Lappé agrees that climate change is with us and it's too late to prevent suffering. But it's not too late for life, she contends. "When facing staggering setbacks—from the Black Death of the fourteenth century to world wars killing nearly 100 million people in the twentieth— most human beings don't end up ruing life. What makes us miserable isn't a big challenge. It's feeling futile, alone, confused, discounted—in a word, powerless. By contrast, those confronting daunting obstacles, but joined with others in common purpose, have to me often seemed to be the most alive."[51]

I wrote this book as a hopeful wake-up call to Canadians, to give them the information and inspiration to act. We need to capture Canadians' imaginations with the idea of a Canadian energy security and conservation strategy. Our current society is not the best humans are capable of. We have to cast off discouragement and join with tens of thousands of Canadians and millions beyond our borders who are already building new, low-carbon societies, ones that are more socially just and far more satisfying. We have a new world to create. The time to start is now.

Glossary

Big Oil

The world's six largest privately owned oil and natural gas corporations: BP, Chevron, ExxonMobil, Royal Dutch Shell, Total and ConocoPhillips.

Bitumen

A naturally occurring mixture of sand, clay, water and a highly viscous form of petroleum with the consistency of a hockey puck. Contains sulphur, nitrogen, salts, carcinogens, heavy metals and other toxins.

Canadian Association of Petroleum Producers (CAPP)

An advocacy group that lobbies on behalf of most large oil and natural gas corporations in Canada.

Canadian Council of Chief Executives (CCCE)

An advocacy organization of the chief executive officers of most of the 150 or so biggest corporations operating in Canada. Many are foreign controlled.

Cap and trade

A market approach to controlling greenhouse gases. Governments cap the amount of GHGs permitted. Corporations wishing to raise their emissions can buy permits from those requiring fewer permits.

Carbon fuels (also called fossil fuels)

Any fuel whose energy derives principally from the oxidation or burning of carbon. Carbon fuels are either *fossil fuels*—coal, petroleum, natural gas—formed by prehistoric natural processes, such as anaerobic decomposition of buried dead organisms, or *biofuels* derived from the recent growth of organic matter, such as wood.

Carbon tax

A tax levied on the carbon content of fuels.

Conventional oil

Petroleum, or crude oil, extracted from the ground (after drilling) through the natural pressure of the wells, pumping or compression (i.e., not fracked).

Corporate Average Fuel Economy (CAFE)

Regulation of the average fuel economy for a manufacturer's fleet of passenger cars and light trucks sold for a given model year.

Crude oil

Naturally occurring, unrefined petroleum composed of hydrocarbon deposits. Crude oil can be refined into gasoline, diesel, and jet fuels. Light crude oil has a low density and flows freely at room temperature. Heavy crude oil has an asphaltic, dense, viscous nature (similar to molasses). Synthetic crude oil is the output from a bitumen, extra heavy oil or shale oil upgrader to make it transportable. It's an intermediate product that is typically further refined into usable products.

Drawdown

A decline in well pressure over time as oil is withdrawn from a well. A measure of the well's production rate.

Dutch Disease

The harmful consequences of a resource boom that drives up the value of a country's currency and makes its manufacturing sector uncompetitive.

Ecological footprint

A measure of human demand on the earth's ecosystems contrasted with its ecological capacity to regenerate.

Economic rent

The increased value, the difference between price and costs, collected by the owner of a resource. The gain in value by virtue of ownership (rather than by production) of land, resources or other natural capabilities. Natural economic rents on public lands are non-renewable depletion charges, part of a country's, province's or First Nation's patrimony.

Energy efficiency

Use of less energy to provide the same service.

Energy Information Administration (EIA) (US)

The main US agency responsible for collecting, analyzing and disseminating energy information. Part of the US Department of Energy.

Energy satellite

A formally independent country that is dominated by a more powerful country, through deep ideological allegiance and economic dependence. An outside power gains first access to the satellite's domestic energy resources.

Energy security

The guarantee of enough energy at affordable prices for everyone in a political community. Immunity from international energy supply disruptions.

Energy superpower

A world leader in energy output and energy reserves that can influence events by projecting economic, military, political and cultural power on a world scale. Energy superpowers put their energy interests above those of other countries and have little energy dependence on other countries or outside influences.

Environmental security

Protecting individuals, communities and countries from environmental risks and changes. Includes food and water security and sovereignty.

Filling costs (of strategic petroleum reserves)

The cost of buying the oil or gasoline to put into the reserves.

Fossil fuels (see carbon fuels)

Fracking (see hydraulic fracturing)

Fuel poverty

When a household cannot afford to adequately warm its dwelling at a reasonable cost, given its income.

Greenhouse gases (GHGS)

Any of the atmospheric gases that contribute to solar warming of the earth's surface—primarily water vapour, carbon dioxide, methane, nitrous oxide and ozone. Human-caused GHGS come mainly from combusting carbon fuels—wood, coal, oil and natural gas.

Hydraulic fracturing (fracking)

Fracturing rock with a pressurized liquid, usually water mixed with sand, ceramic and about 750 chemicals, such as hydrochloric acid and peroxodisulfates, to create small fractures to allow natural gas or oil to migrate to a well. Grains of sand hold the small fractures open. First commercially used in 1949. Companies often don't disclose the chemicals used.

International Energy Agency (IEA)

Set up in 1974 to protect rich oil-consuming countries from the power of OPEC (Organization of Petroleum Exporting Countries) over the price and supply of oil in the world.

Kyoto Accord

A 1997 protocol of the United Nations Framework Convention on Climate Change. An international treaty that set binding obligations on industrialized countries to reduce emissions of greenhouse gases to below 1990 levels, based on the premise that global warming exists and is human caused. Countries in the Global South were exempt from emission reductions because they did not appreciably contribute to GHGs until recently. One hundred and ninety-two states (including all UN members except Andorra, Canada, South Sudan and the United States) are party to the convention. Canada ratified it but withdrew in 2011. A second round of climate negotiations has taken place subsequent to the Kyoto Accord. The next one will be held in Paris in late 2015.

Liquified natural gas (LNG)

Natural gas (mainly methane) that has been converted to liquid form for ease of storage or transport. It takes up about 1/600th the volume of natural gas in the gaseous state.

Low-carbon fuel standard (LCFS)

A regulation to decrease the carbon intensity of transportation fuels, such as gasoline or diesel. Conventional petroleum fuels, including gasoline, are used as the standard for comparison. California enacted the world's first LCFS in 2007, but reductions in carbon intensity will start only in 2016.

National Energy Board (NEB) (Canada)

Established in 1959 to ensure that Canadian energy be developed in the national public interest, the NEB regulates pipelines, energy development and trade. It is formally accountable to Parliament through the Minister of Natural Resources.

National Energy Program (NEP) (Canada)

The energy program of the Pierre Trudeau government 1980 to 1985 that aimed to gain Canadian control over energy through security of supply and independence, substantially raise Canadian ownership and control over the energy industry, maintain a lower Canadian oil price, and significantly shift energy revenues from Alberta to the federal government.

National energy quotas (NEQS)

An energy rationing system for dealing with peak oil and climate change. NEQS apply only to energy purchases and are designed for the national scale. They are called *tradable energy quotas* in Britain and Europe.

Nationalization

Taking a privately owned company into public ownership by a government at the national, sub-national (e.g., province or state) or municipal level.

Natural gas liquids
Liquids from natural gas that are separated from the gaseous state at a field facility or gas processing plant. They include ethane, propane, butane, isobutane, and pentane and natural gasoline (pentanes plus). Gasoline can be made from the latter two.

Neoliberalism
An economic doctrine, developed by Friedrich Hayek, Ayn Rand and Milton Friedman, which claims that extensive economic liberalization including privatization, fiscal austerity, deregulation, corporate rights agreements, and reductions in government spending, will increase economic growth and freedom by giving primacy to the private sector and corporations.

North American Free Trade Agreement (NAFTA)
The 1994 international agreement between the us, Mexico and Canada for closer economic and energy integration. nafta removes the national barriers to corporate entry.

Oil play
Oil fields or prospects in the same region and subject to the same geological conditions.

Oil sands
The terms *oil sands* and *tar sands* have been used interchangeably to refer to bitumen. In the mid-1990s, oil corporations and the Alberta and federal governments began a public relations campaign to substitute "oil sands" for the dirtier-sounding "tar sands."

Organization of Petroleum Exporting Countries (OPEC)
A permanent, international organization of 12 petroleum producing countries (Algeria, Angola, Ecuador, Iran, Iraq, Kuwait, Libya, Nigeria, Qatar, Saudi Arabia, the United Arab Emirates and Venezuela) that aims to lessen the power of giant oil corporations in the global North, bring greater revenues to the national owners of energy resources, and enhance the sovereignty of oil exporting countries in the global South.

Peak oil
The time when the maximum rate of petroleum extraction is reached nationally or globally.

Petro-elites
Include the oil and natural gas corporations (privately or publicly

owned), and other corporations, media and government officials or
bodies that support the petroleum corporations' interests. If government
agencies (e.g., the National Energy Board) that are supposed to regulate
oil and natural gas corporations are captured by the corporations they
are supposed to oversee, they are among the petro-elites.

Proportionality
A nafta rule requiring member countries to continue to make available
for export the same proportion of total energy supply that it has over the
previous three years. Its main aim has been to ensure us access to Canadian energy resources. Mexico gained an exemption from proportionality, which also releases the United States from an obligation to export its
natural gas to Mexico.

Public interest ownership
Entities that must act in the public interest, as defined by new rules of
incorporation. They must be not-for-profit and could be owned by a level
of government (federal, provincial, local or indigenous) or be employee-
owned co-operatives.

Refining of oil
The chemical processes used to transform crude oil into useful products,
such as propane, gasoline, kerosene, jet fuel, diesel oil and furnace oil.

Renewable energy
Derived from natural processes that are replenished continually.
Renewable energy sources include solar, wind, ocean, hydropower,
biomass, geothermal and biofuels.

Rents (see economic rents)

Royalties
Usage-based payment made by the "licensee" to the "licensor" for the
right to ongoing use of an asset. Royalties are not taxes, even when
charged by governments. Private woodlot owners, musicians and
governments collect royalties as owners of an asset. Royalties are one
way governments can capture *economic rents* on resources they own.
Leases, government ownership, ecological charges and corporate taxes
are other ways.

Sands
The term used in this book for bitumen, oil sands and tar sands.

Shale oil

Shale oil, also called kerogen oil or oil-shale oil, is an unconventional oil produced from oil shale rock fragments. It can be converted into synthetic oil and natural gas. Shale oil is usually upgraded into refinery oil by adding hydrogen and removing sulphur and nitrogen.

Strategic petroleum reserves (SPRS)

Emergency storage pools of oil to be released during oil supply disruptions. The standard is to have ninety days' worth of net imports. SPRS usually contain crude oil but can also use refined oil.

Tar sands (see oil sands)

Tight oil

Light crude oil contained in formations of low permeability, often shale or tight sandstone. Tight oil requires fracking and often horizontal drilling.

Tradable energy quotas

Developed in Britain by David Fleming, they were originally called "domestic tradable quotas." (See *national energy quotas*.)

Transfer pricing

Setting the price of goods and services between controlled entities within a firm, often across international borders. Not an arms-length transaction. Transfer pricing is a major tool for corporate tax avoidance.

Unconventional oil

Produced or extracted using non-conventional techniques. Unconventional oil is harder and more costly to produce than conventional oil, requiring more energy, and it usually emits more carbon. It includes extra heavy oil, oil shale, bitumen, biofuels, and conversion of coal to liquid and natural gas to liquid. Unconventional oil is not a fixed category. Resources considered unconventional can later be deemed conventional.

Upgrader

A facility that upgrades bitumen into synthetic crude oil.

Endnotes

Preface
1 "Tories Storm Out of Meeting on Sharing Energy with US," *Ottawa Citizen,* May 11, 2007, A3.
2 Cy Gonick, "Think Nationally Act Locally," *Canadian Dimension,* no. 3 (June/July 1995), 4.

Chapter 1
1 Lita Epstein et al, *The Complete Idiot's Guide to the Politics of Oil,* (Indianapolis: Penguin, 2003), 32–3.
2 Ibid., 31–2.
3 *IEA Response System for Oil Supply Emergencies,* (Paris: IEA, 2012), 11, www.iea.org.
4 Stephen Harper, interview by Peter Mansbridge, "Peter Mansbridge Talks with Stephen Harper," transcript, *CBC News,* January 16, 2012, www.cbc.ca.
5 Meagan Clark, "Why The US Bans Crude Oil Exports: A Brief History," *International Business Times,* March 20, 2014.
6 Leonardo Maugeri, "Oil: The Next Revolution," (Discussion Paper 2012-10, Belfer Center for Science and International Affairs, Harvard Kennedy School, June 2012), 57.
7 Karen Kleiss, "Redford rejects criticism of proposed Canadian energy strategy," *Edmonton Journal,* May 10, 2012.
8 David Thompson, "Toward an Energy Security Strategy for Canada," (Discussion Paper, Edmonton: Parkland Institute, December 2005), http://parklandinstitute.ca/research/tag/energy+policy/.
9 Alexander Panetta, "Is Barack Obama Telling the Truth about the Keystone XL pipeline?," *CBC News,* November 18, 2014.
10 "Basic Statistics," Canadian Association of Petroleum Producers, www.capp.ca.
11 "Refining and Wholesale Markets," Consumers Council of Canada, www.consumerscouncil.com. Canada is a net exporter of gasoline, but imports some too: 4.6 billion litres in 2007, mostly into Quebec. Much of this is subsequently shipped to Ontario.
12 Matt Simmons, interview by Anna Maria Tremonti, *Current,* CBC, February 6, 2008, www.cbc.ca.
13 Elizabeth May, "Pipelines to the East?," *Island Tides,* Elizabeth May's website, April 25, 2013, www.elizabethmaymp.ca.
14 Tom Whipple, "The Peak Oil Crisis: A Reality Check," *Falls Church News-Press,* October 29, 2014; "Countries Oil Consumption," Energy Information Administration (US), accessed December 1, 2014, www.eia.gov. Gasoline for American drivers is also made from US natural gas liquids, which have also risen.
15 J. David Hughes, "A Reality Check on the Shale Revolution," *Nature* 494, (2013), 307–8, doi: 10.1038/494307a.
16 Louis Sahagun, "US Officials Cut Estimates of Recoverable Monterey Shale Oil by 96 Percent," *Los Angeles Times,* May 20, 2014.
17 "Overview Data for United States," EIA (US), accessed May 21, 2013, www.eia.gov.
18 Richard Heinberg, "World Crude Oil Production 2001–2011," Table based on data from the EIA, emailed to author on April 4, 2013. World crude oil production was 73,802,000 barrels a day in 2005, 72,361,000 bpd in 2009 and 74,060,000 bpd in 2011.
19 If emissions are limited to 565 more Gt CO2, scientists calculate there is a 20% chance global warming will exceed 2 degrees Celsius by 2050.
20 *Unburnable Carbon: Are the World's Financial Markets Carrying a Carbon Bubble?,* (London: Carbon Tracker Initiative, November 2011), www.carbontracker.org/library/#carbon-bubble.
21 William Marsden, "Stephen Harper Touts Keystone XL Pipeline in New York, Downplays Oilsands Emissions," *Postmedia News,* May 16, 2013, www.canada.com.
22 Will Campbell, "Oil, Gas Sector is Now Canada's Biggest Source of Greenhouse Gases," *Toronto Star,* April 12, 2014.
23 Greenhouse Gas (GHG) Emissions," How Canada Performs, The Conference Board of Canada, January 2013, www.conferenceboard.ca/hcp/details/environment/greenhouse-gas-emissions.aspx.
24 Edgar G. Hertwich and Glen P. Peters, "Carbon Footprint of Nations: A Global, Trade-Linked Analysis," *Environmental Science and Technology* 43, no. 16 (2009), 6416.
25 Anders Hayden, *When Green Growth Is Not Enough: Climate Change, Ecological Modernization, and Sufficiency,* (Montreal: McGill-Queen's, 2014), 43; Private Member's Bill: Climate Change Accountability Act, Parliament of Canada, 40ᵗʰ Parliament, 3rd Session, March 3, 2010 - March 26, 2011, www.parl.gc.ca. The House of Commons passed the bills under different numbers in 2008, 2009 and 2010 but they were blocked by the 2008 election, the 2009 prorogation and the unelected Senate in 2011 from becoming law.

[26] *Making Sweden an OIL-FREE Society*, (Stockholm: Commission on Oil Independence, 2006), 10–11, www.government.se. For example, in 2005, Sweden's Social Democratic government audaciously pledged to break Sweden's oil dependence by 2020. Unfortunately, the Moderate (conservative) led governments from 2006 to 2014 did not vigorously pursue the pledge; Tim Barber and Eugene Lang, "Gov't Silence on Oilsands an NEP Legacy," *Edmonton Journal*, Nov 1, 2010, A17; David Suzuki and Faisal Moola, "Let's Dare to Consider a National Energy Plan for Canada," *Science Matters Blog*, David Suzuki Foundation, 2010, www.davidsuzuki.org.

[27] Romina Maurino, "Canada Needs National Energy Strategy to Remain Industry Leader: Enbridge CEO," *Canadian Press*, March 23, 2006.

[28] Dr. Roger Gibbins and Dr. Kari Roberts, *Canada's Power Play: The Case for a Canadian Energy Strategy for a Carbon-Constrained World*, (Calgary: Canada West Foundation, September 2008), www.cwf.ca

[29] Karen Kleiss, "Provinces Look to Find Common Ground on Energy Strategy," *Financial Post*, January 16, 2012.

[30] Roger Gibbins, "Ontario Coming to the Table: Creating a Canadian Energy Strategy," Canada West Foundation, March 6, 2012, www.cwf.ca.

[31] Andrew Nikiforuk, "Bruce Carson Scandal Greased by Harper's Oil Sands Agenda," *Tyee*, April 27, 2011, thetyee.ca; "Ex-Harper aide left Calgary school with bill," *Edmonton Journal*, October 13, 2011, A11; Geoff Dembicki, "Bruce Carson's Fingerprints on 'National Energy Strategy,'" *Tyee*, July 21, 2011, thetyee.ca.

[32] Jack Mintz, "National Energy Strategy would Be Highly Dangerous," *Financial Post*, May 4, 2012.

[33] Stephen Harper, quoted in "Canadian Energy Strategy: Stephen Harper Keen to Learn Details of Alberta's Plan," Canadian Press, *Huffington Post*, last updated March 6, 2012, www.huffingtonpost.ca.

[34] Marci McDonald, *Yankee Doodle Dandy: Brian Mulroney and the American Agenda*, (Toronto: Stoddart, 1995), 221–2; "Article 605," *North American Free Trade Agreement*, accessed June 15, 2012, www.international.gc.ca; The proportion of Canadian natural gas output that is exported to the US is dropping as US domestic supplies, mainly from shale, rise.

[35] Cyndee Todgham Cherniak, interview by author, August 21, 2008. Cherniak, a Toronto trade lawyer, told the author that she doesn't know who came up with the idea of proportionality, but never heard it was pushed by Canadians. She suspects the US demanded it because of their reliance on outside energy.

[36] Frances Russell, "Free Trade: Destruction of Canadian Sovereignty," *Outlook*, Jan/Feb 2008, www.vcn.bc.ca.

[37] "Gordon Brown's Speech on Climate Change," *Guardian*, March 15, 2005, www.theguardian.com.

[38] Gordon Laxer, Diana Gibson, and David Thompson, *Toward an Energy Security Strategy for Canada: A Discussion Paper*, (Edmonton: Parkland Institute, 2005), www.parklandinstitute.ca.

[39] A prominent Canadian official who was involved in the FTA negotiations, and who wants to remain anonymous, said as much in a private communication.

[40] Paul Wells, "Harper's Sleepy Majority," *Maclean's*, January 6, 2012, www.macleans.ca.

[41] Becket Adams, "Canadian PM Blasts Obama on Keystone: The US Is an Unreliable Energy Partner," *Blaze*, April 4, 2012, www.theblaze.com.

[42] *Transporting Crude Oil by Rail in Canada*, CAPP, March 2014, p. 6, fig. 2-2, www.capp.ca.

[43] *World Energy Outlook (WEO) 2013*, (Paris: International Energy Agency, 2013), 23, www.iea.org.

[44] Frank McKenna, "Don't Put All Our Oil in the US Basket," *Financial Post*, November 28, 2011, business.financialpost.com.

[45] Robyn Allan, "Canadian Oil Producers' Crocodile Tears: Two Reasons Why Claims that Pipeline Resistance Hurts Their Bottom Line Are, Well, Crude at Best," *Tyee*, January 29, 2013, thetyee.ca.

[46] Robyn Allan, *Bitumen's Deep Discount Deception and Canada's Pipeline Mania: an Economic and Financial Analysis*, April 2, 2013, 4, www.robynallan.com.

[47] Ian Urquhart, "Bitumen Bubble Is More Bluff than Reality," *Calgary Herald*, March 26, 2013, www.calgaryherald.com.

[48] Robyn Allan, "Oil Sands 'Money Left on the Table' and More Myths," *Tyee*, April 11, 2013, thetyee.ca.

[49] Frank McKenna, "A Pipeline from Coast to Coast," *Globe and Mail*, June 18, 2012, A13.

[50] Derek Burney and Eddie Goldenberg, "Shipping Oil to Asia? The Route's East, Not West," *Globe and Mail*, December 13, 2011, www.theglobeandmail.com.

[51] Marc Lee and Amanda Card, *Peddling GHGs: What Is the Carbon Footprint of Canada's Fossil Fuel Exports?*, (Ottawa: Canadian Centre for Policy Alternatives, 2011), www.policyalternatives.ca.

[52] *Canada's Emissions Trends*, (Ottawa: Environment Canada, 2014), 25, www.ec.gc.ca.

[53] Prius Chat, "Pew Poll Finds Strong Support for 56MPG Fuel Economy Standards," July 28, 2011, www.priuschat.com/threads/pew-poll-finds-strong-public-support-for-56-mpg-fuel-economy-standard.96436.

[54] John Kerry, "Three New Bold Ideas for Energy Independence and Global Climate Change," Speech, Faneuil Hall, Boston, Massachusetts, June 26, 2006.

[55] President George W. Bush, "State of the Union Address," January 31, 2006, accessed July 24, 2007, georgewbush-whitehouse.archives.gov; T. Boone Pickens, John D. Podesta and Harry Reid, "Solving Our Nation's Energy Problems," *Chicago Tribune*, February 24, 2009, www.chicagotribune.com; John Kingston, "US Net Oil Import Dependence Drops Another Notch," *The Barrel Blog*, Platts website, June 29, 2012, www.platts.com.

[56] "Oil Net Imports Have Declined Since 2011, with Their Value Falling Slower than Volume," EIA (US), February 25, 2014, www.eia.gov.

[57] *National Energy Policy*, Report, National Energy Policy Development Group (US), May 2001, www.wtrg.com; *Energy Independence and Security Act*, US Environmental Protection Agency, 2007, www.epa.gov.

[58] US, "Driving Efficiency: Cutting Costs for Families at the Pump and Slashing Dependence on Oil," July 2011, p. 1, whitehouse.gov.

[59] Ibid., 7.

[60] "Key Trends Impacting Natural Gas Prices in the US," NASDAQ website, January 3, 2014, www.nasdaq.com.

[61] "Country Comparison: Natural Gas Exports," *World Factbook*, CIA, 2012 estimate, accessed December 2, 2014, www.cia.gov; "Natural Gas Reserves and Production," Oil and Gas Info, accessed January 7, 2014, www.oilandgasinfo.ca; Bill Powers, *Cold, Hungry and in the Dark*, (Gabriola Island, BC: New Society Publishers, 2013), 33.

[62] Ibid., 172; "Canada's Energy Future: Energy Supply and Demand Projections to 2035," National Energy Board (Canada), accessed Jan 8, 2014, www.neb-one.gc.ca. The NEB expects Canadian natural gas output to rise.

[63] David Hughes, "Canadian Gas Exports Threaten Energy Security," *Watershed Sentinel*, November–December 2011, 31–5.

[64] Powers, *Cold, Hungry and in the Dark*, 172.

[65] Susan Eaton, "Marginal Carbons Key to Prosperity: Dinning," Business Edge 4, no. 30 (2004).

[66] Roberts, *End of Oil*, 218–219; Robert L. Hirsch, Roger Bezdek and Robert Wendling, *Peaking of World Oil Production: Impacts, Mitigation, & Risk Management (Hirsch Report)*, (Washington, DC: US Dept of Energy, 2005), 20. This report gives slightly different figures: a 13% drop from 1973 to 1983.

[67] Dan Plesch, "New Energy for Global Security," in *Britain's Energy Future: Securing the 'Home Front,'* Dan Plesch, Greg Austin and Fiona Grant, (London: The Foreign Policy Centre, 2005), 7–8, fpc.org.uk; Ibid., 9; "Iraq Overview," EIA (US), January 30, 2015; "Iraq Crude Oil Production By Year," chart based on EIA data, Index Omundi, accessed April 14, 2015, www.indexmundi.com. Iraq's oil output has grown considerably since 2011 though. The Islamic State of Iraq and the Levant ISIL did not reduce Iraq's oil exports much.

[68] Rusty G. Braziel, "I'll Be Back—Will California's Low-Carbon Rule Terminate Refineries there?," RBN Energy LLC, Feb 23, 2015, rbnenergy.com.

[69] *Energy Independence and Security Act of 2007*, United States Code, www.gpo.gov.

[70] President Barak Obama, interview by Peter Mansbridge, *National*, CBC News, February 17, 2009, www.cbc.ca.

[71] Colin Horgan, "Heads in the Tar Sands," *Guardian* (London), February 20, 2009. www.theguardian.com

[72] "How Much Petroleum does the United States Import and from Where?," EIA (US), November 3, 2014. In 2012, the US got 20% of its petroleum products imports from Persian Gulf states and another 9% from Venezuela.

[73] Larry Pratt, "Pipelines and Pipe Dreams: Energy and Continental Security," in *Whose Canada? Continental Integration, Fortress North America, and the Corporate Agenda*, ed. Ricardo Grinspun and Yasmine Shamsie, (Montreal: McGill-Queen's University Press, 2007).

[74] William Marsden, *Stupid to the Last Drop*, (Toronto: Random House, 2007), 75.

[75] Jeremy van Loon, "Canadians Overestimate Economic Impact of Oil Sands: Survey," *Bloomberg*, July 4, 2014.

[76] Dick Cheney, "Speech at the Institute of Petroleum Autumn lunch", Full Text, November 15, 1999, posted on Get Real List website, www.getreallist.com.

[77] The Canadian Press, "Canada bound for oil's 'front ranks': CIBC," *Business Edge* 7, no. 19 (2007), www.businessedge.ca. Rubin chose a wide percentage spread to take into account of "your view of the investment climate in Kazakhstan and Nigeria."

[78] Diana Gibson, *Selling Albertans Short: Alberta's Royalty Review Panel Fails the Public Interest*, (Edmonton: Parkland Institute, 2007), 5, www.ualberta.ca.

[79] Claudia Cattaneo, "PetroChina Bids to Help Build $5.5-billion Northern Gateway Pipeline," *Financial Post*, March 28, 2012, business.financialpost.com. Enbridge called this story "speculation in the extreme"; "Who Owns Our Oil Sands? Foreign corporations stake their claims to our resources," *Union Magazine*, Alberta Federation of Labour, April 22, 2011, www.afl.org; "South Korea Buys into Alberta Oil Sands," *Engineering and Mining Journal* 207, no. 7 (2006), 6; "Abu Dhabi Spending Billions in Alberta," *Arab American News*, December 1, 2007, accessed July 16, 2008, www.arabamericannews.com.

[80] Theophilus Argitis and Jeremy van Loon, "Harper Says Canada Needs Foreign Investment Discretion," *Bloomberg*, November 8, 2013, www.bloomberg.com.

[81] "Who Owns Our Oil Sands?," *Union Magazine*, www.afl.org.

[82] *How Canadians Perceive Gas Price Hikes*, Survey, (Montreal: Leger Marketing, 2005), 1–11, www.leger360.com. A higher percentage wanted to nationalize the oil.

[83] "Global Greenhouse Gas Emissions Data," EPA (US), Sept 9, 2013, www.epa.gov; *National Inventory Report 1990–2012*, Executive Summary, (Ottawa: Environment Canada, 2014), 1; "Greenhouse Gas Emissions," *Alberta's Oil Sands*, Alberta Government, accessed Jul 2, 2014, www.oilsands.alberta.ca.

Chapter 2

[1] Harold Innis, "Great Britain, The United States and Canada," in *Essays in Canadian Economic History*, (Toronto: University of Toronto Press, 1956), 405.

[2] The President's Materials Policy Commission (Paley Commission), *Resources for Freedom; a Report to the President*, (Washington, U.S. Govt. Print Office, 1952).

[3] Paley Commission, *Resources for Freedom*, 7.

[4] David S. Painter, *Oil and the American Century: The Political Economy of US Foreign Oil Policy, 1941-1954*. (Baltimore: The John Hopkins University Press, 1986), 16.

[5] Ibid.

[6] Although the boycott of Iran temporarily boosted US imports of Saudi oil, over the past decade, the latter has been slowly declining, to the point where China now gets more Saudi oil than the US does.

[7] Painter, *Oil and the American Century*, 16.

[8] Jimmy Carter, "*The President's Proposed Energy Policy*," Speech, April 18, 1977, transcript posted on Minnesotans For Sustainability, www.mnforsustain.org.

[9] Paul Roberts, *The End of Oil: On the Edge of a Perilous New World*, (Boston: Houghton Mifflin, 2004), 153.

[10] The American Presidency Project, Jan 23, 1980, www.presidency.ucsb.edu/ws/?pid=33079.

[11] Innis, "Great Britain, the United States and Canada," *Canadian Economic History*, 407.

[12] George Grant, *Lament for a Nation*, (Toronto: McClelland & Stewart, 1965), 43; James Laxer, *Oil and Gas*, (Toronto: James Lorimer, 1983), 8; Stephen Clarkson, *Canada and the Reagan Challenge*, (Toronto: Lorimer, 1985), 58.

[13] This was English Canada's national vision, not Quebec's.

[14] G. Bruce Doern and Glen Toner, *The Politics of Energy: The Development and Implementation of the NEP.* (Toronto: Methuen, 1985), 7.

[15] J.M. Stopford, Susan George and John S. Henley, *Rival States. Rival Firms*, (Cambridge: Cambridge University Press, 1991), 15. 336 transnationals in the world were nationalized and de-globalized from 1970 to 1975.

[16] Larry Pratt, "Energy, Regionalism and Canadian Nationalism," *Newfoundland Studies* 1, no. 2 (1985), 180; Doern and Toner, *Politics of Energy*, 107. Canadian Petroleum Association's confidential poll; Ibid., 108; At first all the revenue from the Canadian ownership account (the gas tax) went to Petro-Canada to finance the buyout of PetroFina. Later, the funds

went to general revenues; John Erik Fossum, *Oil, the State and Federalism: The Rise and Demise of Petro-Canada as a Statist Impulse*, (Toronto: University of Toronto Press, 1997), 176, 214. Canadian ownership was defined as over 39 per cent of a company's shares. Canadian control as opposed to ownership rose to high of 43% in 1985.

17 *US Imports from Canada of crude oil and petroleum products 1973-2014*, US Energy Information Administration, www.eia.gov. Imports were 52.1 million cubic metres in 1973. By 1983, these had fallen to 14.6 million cubic metres.

18 Fossum, *Oil, the State and Federalism,* 125.

19 "Preface," *National Energy Program*, (Ottawa: Canada, 1980), 1–2.

20 Doern and Toner, *Politics of Energy*, 20.

21 Denise Harrington, "Who Are the Separatists?," in *Western Separatism: The Myths, Realities & Dangers*, eds. L. Pratt and G. Stevenson, (Edmonton: Hurtig, 1981), 29–30.

22 Larry Pratt, "Energy, Regionalism and Canadian Nationalism," *Newfoundland Studies* 1, no. 2 (1985), 178.

23 Doern and Toner, *Politics of Energy*, 266–75.

24 Pratt, "Energy, Regionalism and Canadian Nationalism," 57.

25 Stephen Clarkson, *Canada and the Reagan Challenge,* (Toronto: Lorimer, 1985), 35.

26 Pratt, "Energy, Regionalism," 191-2.

27 Godfrey Budd, "Global Recession Sidelined the NEP's Effects as the Provincial Economy Slowed," *Alberta Construction Magazine*, March/April 2009, 53.

28 The international price of oil hit $40 US a barrel in 1981 and fell below $10 a barrel by 1986. This was a period of high inflation, so the price of oil fell much more in constant dollars.

29 Dan Crawford, "The Plan Is…There Is No Plan," *Republic News*, accessed October 23, 2009.

30 Based on an extended phone conversation with the author, December 4, 2013.

31 Premier Ed Stelmach, "The Way Forward," Speech, October 14, 2009; Hugh McCullum, *Fuelling Fortress America*, (CCPA, Polaris Institute and Parkland Institute, 2006), 6; Canadian Press, "Energy Minister Says Alta. Has Enough Oil to Last Centuries," *The Evening News* (New Glasgow) March 8, 2006, A10; Gary Lunn, letter to Post Carbon Toronto, March 7, 2008, www.postcarbontoronto.org/results-of-activities/correspondence/. The exporting explanation made no sense as a reason to reject SPRs for Canada. But, it's part of Mr. Lunn's letter.

32 "CANSIM table 126-0001: Supply and disposition of crude oil and equivalent," *CANSIM* (database), Statistics Canada, www.statcan.gc.ca. 40% in 2012. 58% in 2011. 55% in 2010. 61% in 2009 & 2008. Calculations by Tanya Whyte. Without access to the refineries' books (for crude inputs and production sales), it is impossible to know how much Newfoundland oil was consumed in Canada.

33 "Table 4.1," *Energy Statistics Handbook*, Statistics Canada, 2013.

34 *Oil Supply Security: Emergency Response of IEA Countries*, (Paris: IEA, 2007), www.iea.org.

35 Ibid., 28.

36 Gary Lunn, letter to Post Carbon Toronto.

37 Council of Canadians et al, *TransCanada's Energy East: An Export Pipeline, NOT for Domestic Gain*, March 2014.

38 "Canada 2010 Update," *Oil & Gas Security: Emergency Response of IEA Countries*, (Paris: IEA, 2010), 3.

39 "Table 126-0001: Supply and disposition of crude oil and equivalent" and "134-0004: Refinery supply of crude oil and equivalent," *CANSIM* (database), Statistics Canada, September 2013. In 2010 and 2011, Nfld's crude oil production was sufficient. In 2012, it could have met 97% of refiners' demand. Atlantic Canadian refiners are net exporters of refined oil though, so that if they were re-geared to meet only Atlantic Canadian consumers demand, Nfld crude oil output is sufficient.

40 Laura Neilson, "Ice Storm 1998," The Canadian Encyclopedia, www.thecanadianencyclopedia.com.

41 "Heating Oil Runs Low in Cape Breton," *CBC News*, December 18, 2007, www.cbc.ca.

42 "Ontario's Gasoline Crisis," *The EN-PRO Advisor*, no. 98, February 27, 2007; "Petro-Canada Stations Running Dry across Prairies," *CBC News*, May 2, 2013, www.cbc.ca; Lauren Krugel, "Glitch at Petro-Can's Edmonton Refinery Leads to Gas Shortages in Alta.," *Cape Breton Post,* December 8, 2008, www.capebretonpost.com; The Alberta NDP are talking about doing more upgrading of Sands oil in Alberta. They are emphasizing upgraders more than refineries.

[43] Hirsch, Bezdek and Wendling, *Hirsch Report*, 8; Securing America's Future Energy, *Oil Shockwave*, 2005, www.secureenergy.org/projects/oil-shockwave.

[44] Leonardo Maugeri, "Oil: The Next Revolution," (Discussion Paper 2012-10, Belfer Center for Science and International Affairs, Harvard Kennedy School, June 2012), 1; Ibid., 6; Colin Sullivan, "Energy Policy: Has 'Peak Oil' Gone the Way of the Flat Earth Society?," *Energy Wire*, Environment & Energy Publishing, March 22, 2013, 1.

[45] *World Energy Outlook (WEO) 2012*, (Paris: IEA, 2012), 23, www.iea.org; *World Energy Outlook (WEO) 2013*, (Paris: IEA, 2013), 73, www.iea.org; Patti Domm, "US Is on Fast-Track to Energy Independence: Study," CNBC, February 11, 2013; Richard Heinberg, *Snake Oil: How Fracking's False Promise of Plenty imperils our future*, (Santa Rosa, California: Post Carbon Institute, 2013), 8.

[46] *WEO 2012*, 105.

[47] *WEO 2013*, 73.

[48] *WEO 2013*, 24. This assumes the IEA's "New Policies scenario."

[49] Richard A. Kerr, "Are World Oil's Prospects Not Declining All that Fast?," *Science* 337, August 10, 2012, 633; Maugeri was on sabbatical leave from ENI when he wrote the report; Jean Laherrère, "Comments on Maugeri's Oil Revolution," part one, *The Oil Drum*, July 2, 2012, peakoil.com; Ibid.

[50] Heading Out, "Tech Talk—New Energy Report from Harvard Makes Unsupportable Assumptions," *The Oil Drum*, July 2, 2012.

[51] Quotes in the next three paragraphs taken from Terry Macalister, "Key Oil Figures Were Distorted by US Pressure, Says Whistleblower," *Guardian* (London), November 9, 2009, www.theguardian.com.

[52] Hilary Whiteman, "Energy Body Rejects Whistleblower Allegations of Oil Cover Up," *CNN*, November 10, 2009, cnn.com.

[53] *World Energy Outlook 2008*, (Paris: IEA, 2008), 221, www.iea.com. Decline rates were 3.4% for super-giant fields, 6.5% for giant fields, and 10.4% for large fields: the younger the field, the higher the decline rate.

[54] Matthew R. Simmons, "Gauging the Risks of Peak Oil and Gas: Limits to Growth," Speech, ASPO World Conference, Houston, October 18, 2007.

[55] Mikael Höök, Robert Hirsch and Kjell Aleklett, "Giant Oil Field Decline Rates and Their Influence on World Oil Production," *Energy Policy*, April 2013, www.elsevier.com/locate/enpol; Kurt Cobb, "The Decline of the World's Major Oil Fields," *Christian Science Monitor*, April 12, 2013.

[56] Gail Tverberg, "IEA Oil Forecast Unrealistically High; Misses Diminishing Returns," Our Finite World, November 13, 2012, 2, ourfiniteworld.com.

[57] Lynne Kiesling, "Toyota's Long-term Vehicle Strategy: Move beyond Petroleum," Knowledge Problem, June 13, 2008, knowledgeproblem.com/2008/06/13/toyotas_longter/; "Toyota's Jim Lentz Predicts Peak Oil by 2020," Posted by ForaTv, YouTube, Uploaded December 30, 2009, www.youtube.com/watch?v=yVEZE2vM2oM; Christopher Helman, "High Friends in Low Places," *Forbes*, January 26, 2011; *20th World Petroleum Congress*, Official Publication, December 2011, www.world-petroleum.org.

[58] Peter Waldman, "Saudi Arabia was Worried about a Danger Much Bigger than Shale when it Blindsided Oil Markets," *Financial Post*, April 14, 2015.

[59] Countries by Peak Oil Date," True Cost Blog, January 21, 2012, truecostblog.com. Based on BP's annually-released Statistical Review of World Energy. Updated using data released in 2011 that includes production through 2010. Several more countries have passed peak production or are stuck on a production plateau.

[60] Countries by Peak Oil Date," True Cost Blog, January 21, 2012, truecostblog.com. Based on BP's annually-released Statistical Review of World Energy.

[61] Ian Chapman, "The End of Peak Oil? Why this Topic is Still Relevant Despite Recent Denials," *Energy Policy*, 2013, 5.

[62] *Investor Presentation*, PEMEX, April 2012, p 8, www.pemex.com/

[63] "The Oil Crunch: Securing the UK's Energy Future," (First Report of the UK Industry Taskforce on Peak Oil & Energy Security, 2008). Task force member companies include Arup, FirstGroup, Foster and Partners, Scottish and Southern Energy, Solarcentury, Stagecoach Group, Virgin Group, Virgin.

[64] Carlos Manuel Rodriguez and Adam Williams, "Pemex Cuts Crude Goal as Cantarell Wanes Ahead of Energy Debate." *Bloomberg*, July 26, 2013; Allan Wall, "The Future of PEMEX,

Mexico's Petroleum Monopoly, an Ongoing Debate," Mexidata.info, September 2, 2013, www.mexidata.info/id3698.html; Peter Waldman, "Saudi Arabia was Worried About a Danger Much Bigger than Shale When it Blindsided Oil Markets," *Financial Post,* April 14, 2015.

65 Energy Watch Group, *Crude Oil: The Supply Outlook,* (Berlin: Energy Watch Group, 2007), 34; Matthew R. Simmons, *The World's Giant Oilfields,* (Houston: Simmons & Company International, 2007), 1; Robert L. Hirsch, Roger Bezdek and Robert Wendling, *Peaking of World Oil Production: Impacts, Mitigation, & Risk Management,* (Washington, DC: US Dept of Energy, 2005), 11; "Petrobras Starts Production From Lula Northeast Presalt," Oil & Gas Journal, June 17, 2013, www.ogj.com; *WEO 2013,* 28; Conway Irwin, "Onshore US: New Frontier for Sub-Salt?," Breaking Energy, April 15, 2013, breakingenergy.com.

66 Peter Millard, "Brazil Oil Fields May Hold More Than Twice Estimates," *Bloomberg,* January 19, 2011, www.bloomberg.com. In 2011, Hernani Chaves, a professor and former geologist for Petrobras estimated that the total recoverable subsalt reserves in Brazil are 123 billion barrels in total, compared to the government estimate of 50 billion; Chris Nelder, "Oil Majors are Whistling Past the Graveyard," *SmartPlanet Blog,* March 20, 2013, hosted at zdnet.com; Jack Santa Barbara, "Peak Oil and Alternative Energy," in *Energy Security and Climate Change,* ed. Cy Gonick, (Halifax: Fernwood Books, 2007), 39.

67 "The Problem of Reliable Data," Planet for Life, planetforlife.com/oilcrisis/oilreserves.html, accessed December 20, 2007; *Crude Oil: The Supply Outlook,* Energy Watch Group, 2007, 33, www.energywatchgroup.de. Some governments in OPEC argued that past reserve assess ments were too low, prior to nationalization of the oil industry, because private companies underreported reserves for financial and political reasons.

68 Kurt Cobb, *Christian Science Monitor.*

69 John Vidal, "WikiLeaks Cables: Saudi Arabia Cannot Pump Enough Oil to Keep a Lid on Prices," *Guardian* (London), February 8, 2011, www.theguardian.com; *The Report: Kuwait 2008,* (Oxford Business Group, 2008), 127; Cobb, "Does OPEC Really have 80 Percent of the World's Oil? Maybe Not," *Christian Science Monitor,* September 13, 2013, www.csmonitor.com.

70 David Strahan, *The Last Oil Shock,* (London: John Murray, 2007), 67.

71 *WEO 2012,* 552. Figures are for 2010.

72 Marbek, "Study of Opportunities for Natural Gas in the Transportation Sector," submitted to Natural Resources Canada, March 2010, www.xebecinc.com/pdf/Marbek-NGV-Final-Report-April-2010.pdf.

73 Oil Transit Chokepoints," The Encyclopedia of Earth, May 16, 2013, www.eoearth.org/view/article/155012, sources: Energy Information Administration (US), APEX Tanker Data, Panama Canal Authority, and Suez Canal Authority.

74 "Energy Now: Oil Shockwave," Secure Energy, July 17, 2011, 5:41 min video, www.secureenergy.org/media/video/energy-now-oil-shockwave. Participants, several of them in George W. Bush's administration, concluded that the US just needed to drill more; "Strategic Petroleum Reserve," Government Accountability Office (US), GAO-06-872, August 2006, 5. About 1.5 million barrels of oil supplies were cut from the market; Blake Clayton, "2011 Strategic Petroleum Reserve Release," The Council on Foreign Relations, September 12, 2012, http://i.cfr.org.

75 Mohammed Jamjoom et al, "Yemen Says It Foiled Al Qaeda Plot," *CNN,* August 7, 2013, www.cnn.com; "Militants Blow Up Oil Pipeline in Iraq, Halt Exports," *FirstPost.World,* Mar 26, 2013, www.firstpost.com; "Strait of Hormuz is Chokepoint for 20% of world's oil," 2011 figures, EIA, September 5, 2012, www.eia.gov.

76 "Imports—Mineral fuels," Statistics Canada, Monthly data from June to September 2014.

77 John Blair, *The Control of Oil,* (New York: Vintage, 1976), 275. "The 'Arab embargo' occasioned only limited and temporary dislocations.… in the period of the 'embargo…OPEC output turned out to be virtually the same as in the following year."

78 Joseph Stroupe, "The New World Oil Order," part one, *Asia Times,* November 22, 2006.

79 See http://www.energybulletin.net/17262.html for a list of major North Sea fields and production start dates. Thanks to Kjel Oslund for this point.

80 Stroupe, "The New World Oil Order."

81 Ibid. The oil reserves of Saud Aramco and NIOC are probably overstated. See the point made earlier about OPEC's official reserves. But even if they are half of the official figures, they still greatly dwarf ExxonMobil and Shell's.

82 "The Global Oil Industry Supermajordämmerung," *Economist,* Aug 3, 2013, www.economist.com.

[83] ACIL Tasman, "An Assessment of Australia's Liquid Fuel Vulnerability," Australia Dept of Resources, November 2008; Kathy Leotta, *Implementing the Most Effective Transportation Demand Management; (TDM) Strategies to Quickly Reduce Oil Consumption*, January 2007, postcarboncities.net "Preparing for and Responding to Energy Emergencies," UK Dept of Energy & Climate Change, July 29, 2013, www.gov.uk/preparing-for-and-responding-to-energy-emergencies.

[84] Munroe, "Response to Canada's Plan for Oil Supply Emergencies," (Unpublished Manuscript, summer 2008), 2.

[85] "Preparing for energy emergencies," UK Dept of Energy & Climate Change.

Chapter 3

[1] "Gordon Brown's Speech on Climate Change," *Guardian* (London), March 15, 2005, www.theguardian.com.

[2] *UK Emergency Oil Stocks*, Dept of Energy & Climate Change (UK), May 2009, www.berr.gov.uk/files/file37711.pdf.

[3] Peter Hetherington and Jon Henley, "French-style Fuel Protest Hits Britain," *Guardian* (London), www.theguardian.com, September 9, 2000; "Impact of September 2000 Fuel Price Protests on UK Critical Infrastructure," Public Safety and Emergency Preparedness Canada, January 25, 2005, www.iwar.org.uk/cip/resources/PSEPC/fuel-price-protests.htm; Strahan, *Last Oil Shock*, 13–16.

[4] Phil Hathaway, "The Effect of the Fuel 'protest' on Road Traffic," Dept for Environment, Transport and the Regions (UK), December 3, 2003, www.dft.gov.uk.

[5] Peter Hetherington, "Panic as Oil Blockade Bites," *Guardian* (London), www.theguardian.com, September 12, 2009.

[6] "Impact of September 2000 Fuel Price Protests on UK Critical Infrastructure," Public Safety and Emergency Preparedness Canada, Jan25, 2005, www.iwar.org.uk/cip/resources/PSEPC/fuel-price-protests.htm.

[7] Ibid.

[8] Greg Muttitt and James Marriott, "Pump and Circumstance," *Guardian* (London), October 4, 2000, www.theguardian.com.

[9] Strahan, *Last Oil Shock*, 16.

[10] "Britain's Energy Crisis: How Long till the Lights Go Out?," Economist, August 8, 2009, 11.

[11] "Secret to Sweden's Success with Renewables?," Renewables International, September 9, 2013, www.renewablesinternational.net.

[12] "UK Emergency Oil Stocks," Dept of Energy & Climate Change (UK), Aug 22, 2013, www.gov.uk/government/publications/emergency-oil-stocking-international-obligations; *Strategic Storage and Other Options to Ensure Long-term Gas Security*, (Report to DTI, ILEX Energy Consulting, April 2006); Chris Le Fevre, *Gas Storage in Great Britain*, (Oxford: Oxford Institute for Energy Studies, 2013), 70.

[13] Noel Hulsman, "Canada Lags China on Energy Efficiency," *Insight,* Yahoo Finance Canada, July 13, 2012, http://ca.finance.yahoo.com/. The study was done by the American Council for an Energy-Efficient Economy.

[14] *BP Statistical Review of World Energy 2012*, BP, www.bp.com. Sweden was not part of the 12 country study. Its oil use per capita was 53% that of Canada's.

[15] Jeff Rubin, *Heading for the Exit Lane*, CIBC World Markets, June 26, 2008, 5–6; "Transportation Statistics by Country," International Comparisons, 3rd quarter 2013, internationalcomparisons.org; Ibid.

[16] Simmons died in 2010.

[17] *The Current*, part 3, CBC, February 6, 2008, www.cbc.ca/thecurrent/2008/200802/20080206.html

[18] EIA *2015 Proved Reserves* of Crude Oil, www.eia.com. Proven reserves are those that can be extracted using current technology and prices. According to the EIA Venezuela has 298 billion barrels, Saudi Arabia has 268 billion barrels. Canada 172 bb, and Iran has 158 bb. The size of Saudi Arabia official reserves are suspect. See discussion in Chapter 2.

[19] "Table 134-000: Refinery Supply of Crude Oil & Equivalent," *CANSIM* (database), Statistics Canada, 2012 data.

[20] Giacomo Luciani and François-Loïc Henry, *Strategic Oil Stocks and Security of Supply*, (CEPS Working Document, No. 353, Centre for European Policy Studies, June 2011), 1, www.ceps.eu; Kimberly Amadeo, "Strategic Petroleum Reserve." About.com US Economy. 19 Mar 2013. http://useconomy.about.com/od/suppl1/p/Strategic-Petroleum-Reserve.htm.

"US Imports by Country of Origin: All Countries," EIA (US), accessed December 9, 2013, www.eia.gov. The reserves have enough oil to theoretically supply oil at that rate for 165 days. US oil imports averaged about 10 million bpd in 2013; Kenneth B. Medlock III and Amy Myers Jaffe, *Who Is In the Oil Futures Market and How Has It Changed?* (James Baker III Institute for Public Policy, Rice University, August 26, 2009), 17.

21 Jerry Taylor and Peter Van Doren, "The Case Against the Strategic Petroleum Reserve," *Policy Analysis,* no. 555, November 21, 2008, 1–21, www.cato.org; *Basic Facts about the Strategic Petroleum Reserve,* US National Commission on Energy Policy, 1, www.energycommission.org. Number of days of replacing total US imports (not the 4.4 million b/day maximum release), do not add up to total US reserves. But, numbers cited here come from US government figures; "Strategic Petroleum Reserves," Dept of Energy (US), energy.gov/fe/services/petroleum-reserves.

22 "Petroleum Reserves," Dept of Energy (US), accessed December 9, 2013, www.fe.doe.gov/programs/reserves.

23 "Strategic Petroleum Reserve: Annual Report for 2006," Dept of Energy (US), 31, energy.gov/fe/downloads/historical-spr-annual-reports-congress. Thanks to Kjel Oslund for pointing this option to me.

24 Salt Cavern Reserve," College of Engineering and Engineering Technology, Northern Illinois University, accessed April 24, 2006, www.ceet.niu.edu//techcourses/ungrad/spr_cav.gif.

25 John Blair, *The Control of Oil,* (New York: Vintage, 1976), 275; Thanks to Erin Weir for reminding me of this point. "Member Countries," IEA, accessed Dec 9, 2013, www.iea.org/countries/membercountries/. All 29 IEA countries are also OECD members. The IEA is an autonomous agency linked with the OECD. China and Estonia have applied to join.

26 Ibid.

27 Kristine Kuolt, "Overview of IEA Emergency Procedures and Measures in IEA Member Countries," (Beijing: IEA/China Seminar on Oil Stocks and Emergency Response, 2002), 9–10.

28 *Response System for Oil Supply Emergencies,* (Paris: IEA, 2012), 7, www.iea.org. In 2011, IEA countries held 1.5 bbd of public oil stocks. The US held half of that at 727 mbd. Industry held another 2.6 bbd. Some of that was for commercial purposes. Some was required by governments to be held for national purposes. In addition, non IEA members hold SPRs.

29 Several Eastern European countries are in the EU but not the IEA: Bulgaria, Cyprus, Estonia, Latvia, Lithuania, Malta, Romania and Slovenia.

30 "India Overview," *International Energy Data and Analysis,* EIA (US), March 18, 2013, www.eia.gov; Mandip Singh, "China's Strategic Petroleum Reserves: A Reality Check," Institute for Defence Studies and Analyses, May 21, 2012, www.idsa.in/issuebrief/ChinasStrategicPetroleumReserves_MandipSingh_210512; Enno Harks, "IEA Security of Oil Supply Oil Crisis Mitigation Stocks," accessed January 18, 2008, www.ecn.nl/fileadmin/ecn/units/bs/INDES/indes-eh.pdf.

31 "Market Snapshot: Canadian Crude Oil Imports Decline," National Energy Board, November 13, 2014.

32 "Basic Statistics," CAPP, accessed December 4, 2014, www.capp.ca; "SPR Quick Facts and FAQS," accessed December 4, 2014, DOE (US) www.energy.gov. The US SPR has held 727 million barrels, unchanged since 2009.

33 "Table 134-0004: The Supply and Distribution of Refined Petroleum Products in Canada (monthly)," *CANSIM* (database), Statistics Canada, Sept 2013 data, www.statcan.gc.ca. Cat # 45-004-X.

34 Doug Heath, phone conversation with author, December 4, 2013.

35 R. Glenn Hubbard and Robert J. Weiner, 1985, 'Managing the Strategic Petroleum Reserve: Energy Policy in a Market Setting', Annual Review Energy, Vol 10, p. 519.

36 679,000 bpd in 2011.

37 Line 9 would carry and export US Bakken shale oil too.

38 Harper, interview by Mansbridge (see chap. 1 n. 4).

39 About 550,000 to 600,000 bpd.

40 "Basic statistics," CAPP, www.capp.ca.

41 "Table 126-0001: Supply and disposition of crude oil and equivalent," *CANSIM* (database), Statistics Canada, 2012 data. Just under 200,000 barrels of oil a day.

42 "Basic Statistics," CAPP. Canada produced 1.38 million barrels per day of conventional oil in 2013 and .93 mbpd of mined bitumen and 1.02 mbpd of in situ bitumen for an annual total of 3.33 million bpd. 30 million barrels would take 9.0 days to fill.

[43] Thomas Palley, "Manipulating the Oil Reserve," Economics for Democratic and Open Societies, accessed January 26, 2007, www.thomaspalley.com/?p=65.

[44] Fexix & Scisson, prepared for Energy Department Development Sector, May 6, 1974.

[45] "Image 1002," *Hansard 9110*, January 7, 1974.

[46] *Deep-Well Storage in Salt Caverns–Lambton County*, (Sarnia, Ont: Lambton Industrial Society, Revised 1995), accessed January 12, 2008, www.sarniaenvironment.com/pdf/SLEA-Monograph-L3.pdf.

[47] In 2000, Toronto's average daily water use was 1.2 billion litres. Total volume of Lambton SPRs would equal about ten days' of Toronto's water use. It would likely be within the capacity of Lambton County's watershed. The water would eventually find its way back to the watershed. Thanks to Kjel Oslund for these points.

[48] Charles Hendry, "Speech at the Energy and Utility Forum. House of Commons," October 21, 2011, www.decc.gov.uk/en/content/cms/news/hendry_211010_hendry_211010.aspx.

[49] "Strategic Petroleum Reserve: Quick Facts and Frequently Asked Questions," US Energy Dept, accessed December 23, 2013, www.fossil.energy.gov/programs/reserves/#SPR.

[50] Joseph Doucet, "Should we create a Strategic Petroleum Reserve in Eastern Canada?," (Unpublished, 2008). Doucet, an Enbridge Professor of Energy Policy at the University of Alberta, kindly sent me a draft. His main case against SPRs is cost.

[51] "SPR Quick Facts and FAQS," Dept of Energy (US), accessed December 4, 2014, www.energy.gov. Sale of the Elk Hills field in 1998 to Occidental Petroleum netted an additional $3.65 billion.

[52] "Responsibilities," National Energy Board (Canada), modified July 17, 2012, www.neb-one.gc.ca.

[53] National Energy Board email to author, April 12, 2007.

[54] Rick Munroe, correspondence with author, July 8, 2012. Munroe has had extensive discussions with government officials on this and other Canadian energy security issues.

[55] "Top Countries with Highest Crude Oil Import," World List Namia, accessed December 12, 2013, www.worldlistmania.com; "Population by Rank," *The World Factbook*, CIA, accessed December 12, 2013, www.cia.gov; *Oil Supply Security 2007*, (Paris: IEA, 2007), 106–109; Ibid., 224.

[56] Ibid., 86–92.

Chapter 4

[1] "Alberta's Energy Industry: An Overview," (Alberta, 2013), 2; "Canadian Consumption of Crude Oil & Products," *Statistical Handbook*, CAPP, www.capp.ca/publications-and-statistics/statistics/statistical-handbook. Alberta calculates that the Sands has 168.7 billion barrels of remaining established reserves of crude oil. Canada consumed 547 million barrels in 2012. That would mean 309 years at current rates of use.

[2] Michael A. Levi, *The Canadian Oil Sands: Energy Security vs. Climate Change*, (New York: Council on Foreign Relations, Special Report No. 47, 2009), 25.

[3] Cyndee Todgham Cherniak, interview by author, June 28, 2009.

[4] Richard Heinberg, "Proportionality," *Energy Bulletin / Blog*, February 7, 2008.

[5] Or basic petrochemical.

[6] "Table 4.1," *Energy Statistics Handbook*, Statistics Canada.

[7] "Table 6.1," *Energy Statistics Handbook* 2013, Statistics Canada, www.statcan.gc.ca

[8] "US Imports (Crude Oil) by Country of Origin," EIA (US), accessed September 29, 2013, www.eia.gov; Carlos Salinas De Gortari, *México: The Policy and Politics of Modernization*, trans. Peter Hearn and Patricia Rosas, (Barcelona: Plaza & Janes, 2002), 72; *1917 Constitution of Mexico* (As Amended), accessed July 23, 2008, www.latinamericanstudies.org/mexico/1917-Constitution.htm.

[9] Barbosa, email to author, November 19, 2008.

[10] Red Mexicana de Acción Frente al Libre Comercio.

[11] John Dillon, "The Petroleum Sector Under Continental Integration" in *The Political Economy of North American Free Trade*, eds. Ricardo Grinspun and Max Cameron, (New York: St. Martin's Press, 1993), 326.

[12] Terisa Turner and Diana Gibson, "Back to Hewers of Wood and Drawers of Water," in *Energy, Trade and the Demise of Petrochemicals in Alberta*, (Edmonton: Parkland Institute, 2005), 5. Mexico has since opened exploration licences to foreign oil corporations.

[13] Dillon, "The Petroleum Sector," 326.

[14] Ibid., 3.

15 "US Natural Gas Imports by Country," EIA (US), accessed October 12, 2013, www.eia.gov. US natural gas imports reached their height between 2003 and 2008.

16 "US Dry Natural Gas Production annual," EIA, www.eia.gov. 1973 was the peak year; The 25 year rule was brought in when the National Energy Board was created in 1959.

17 Fast Facts," Good Energy, Bad Energy: Transforming Our Energy System for People and the Planet, Friends of the Earth International, November 8, 2013, www.foejapan.org/en/news/131108.html.

18 Christopher Hatch and Matt Price, *Canada's Toxic Tar Sands: The Most Destructive Project on Earth*, (Toronto: Environmental Defence, February 2008), 21; Gordon Laxer and John Dillon, *Over a Barrel: Exiting from NAFTA's Proportionality Clause*, Parkland Institute and Canadian Centre for Policy Alternatives, May 2008, www.iatp.org; GATT Article XI allows export quotas under certain conditions, while GATT Article XX allows exemptions to free trade.

19 Laxer and Dillon, *Over A Barrel*. The next few paragraphs are partly based on John Dillon's contribution to the report.

20 Cabinet Task Force on Oil Import Control, *The Oil Import Question: A Report on the Relationship of Oil Imports to the National Security*, (Washington, DC: US Government Printing Office, 1970), 94.

21 David S. Painter, *Oil and the American Century*, (Baltimore: The John Hopkins University Press, 1986), chap. 5.

22 Marci McDonald, *Yankee Doodle Dandy: Brian Mulroney and the American Agenda*, (Toronto: Stoddart, 1995), 221–2.

23 Linda McQuaig, *The Quick and the Dead*, (Toronto: Viking, 1991), 173.

24 "Energy Supply Security 2014" (Paris: IEA, 2014) A full response includes stockdraw, demand restraint, fuel-switching, surge oil production, and sharing of available supplies.

25 Dan Plesch, "New Energy for Global Security," in *Britain's Energy Future: Securing the Home Front*, Plesch et al, (London: The Foreign Policy Centre, 2005), 4.

26 Lisa Sanders and William L. Watts, "CNOOC Walks Away from Unocal Deal," *MarketWatch*, August 2, 2005, www.marketwatch.com. In contrast, Canada allowed CNOOC to buy out Nexen Inc., a Canadian owned oil and gas corporation in 2012.

27 *Report of the National Energy Policy Development Group*, National Energy Policy, US Office of the President, May 2001, Chapter 8-8 and 8-9.

28 Richard N. Haass, "Foreword," in Michael A. Levi, *The Canadian Oil Sands: Energy Security vs. Climate Change*, Council on Foreign Relations, Center for Geoeconomic Studies, Council Special Report No. 47, May 2009, vii; Amy Myers Jaffe, *United States and the Middle East: Policies and Dilemmas*, (US National Commission on Energy Policy, Sept, 2003); Ian Urquhart, "Urquhart: Bitumen bubble is more bluff than reality," *Calgary Herald*, March 26, 2013.

29 "Table R5.10," *ST 98-2013*, Alberta's Energy Reserves and Supply/Demand Outlook, (Edmonton: Alberta Energy Regulator, 2013), 5-26; "Natural Gas," *Alberta's Energy Industry: An Overview*, (Edmonton: Alberta Government, 2012). "42% to the United States;" *Enquiry into Reserves and Consumption of Natural Gas in the Province of Alberta*, (Edmonton: Natural Gas Commission, 1949), 122.

30 Paul G. Bradley and G. Campbell Watkins, *Canada and the US: A Seamless Energy Border?*, (CD Howe Institute, no. 178, April 2003), 11, www.cdhowe.org/pdf/commentary_178.pdf. The Alberta Gas Resources Preservation Act, passed in 1949 and amended in 1955, 1970, 1980, 1987, and 2000, promised security for Alberta users of natural gas. "Core consumers" were residential and commercial gas consumers, not the Sands.

31 "Table R5.2," *ST 98-2013*, Alberta Energy Regulator, 5-4.

32 "Talk about Natural Gas," Alberta Energy, www.energy.alberta.ca.

33 "Gas Demand for Alberta Oil Sands to 2020," Ziff Energy Group, August 30, 2011, www.ziffenergy.com. Going from 400 billion cubic feet to 1.1 trillion cubic feet.

34 "Canadian Energy Overview 2011," National Energy Board, 12.

35 *ST 98-2013*, Alberta's Energy Reserves Outlook, AER; *Alberta's Energy Industry*, Alberta Government, 2012. Calculations for 2011 from *ST 98-2013*.

36 McDonald, *Yankee Doodle Dandy*, 221.

37 Maxwell A. Cameron and Brian W. Tomlin, *The Making of NAFTA: How the Deal Was Done*, (Ithaca and London: Cornell University Press, 2000), 230.

38 McDonald, *Yankee Doodle Dandy*, 223.

39 Gordon Ritchie, *Wrestling with an Elephant: The Inside Story of the Canada-US Trade Wars*, (Toronto: Macfarlane, Walter & Ross, 1997), 125–6.

40 Ritchie, "Who's Afraid of NAFTA's Bite?," *Globe and Mail*, February 15, 2005, A21.

41 Russell, "Free Trade," *Outlook* (see chap.1 n.35.).

42 Peter Lougheed, "The Rape of the National Energy Program Will Never Happen Again," in *Free Trade, Free Canada*, ed. Earle Gray, (Woodville, Ont: Canadian Speeches, 1988).

43 Carlos Salinas De Gortari, *México: The Policy and Politics of Modernization*, (Barcelona: Plaza & Janes Editores SA, 2002), 86.

44 *Creating Opportunity: The Liberal Plan for Canada (The Red Book)*, (Ottawa: The Liberal Party of Canada, 1993).

45 Canada's main objections were the US government's undiminished ability to use countervail, anti-dumping and subsidies to block Canadian exports.

46 "PM Warned to Keep Vow on Changing Trade Pact," *Toronto Star*, November 20, 1993. The letter was leaked to the Council of Canadians. Erin Krekoski did excellent research on the 1993 federal election.

47 Clyde Farnsworth, "Chrétien Says He Wants Changes in Trade Accord," *New York Times*, October 28, 1993.

48 Axworthy, "Obama takes the High Road: Will Canada Follow?," *Edmonton Journal*, December 1, 2008, A18.

49 "The Declaration on Energy," *Edmonton Journal*, December 3, 1993.

50 David Vienneau, "PM's Gamble over Energy May Just Work, Experts Say," *Toronto Star*, December 3, 1993, A27. McRae is on the roster of panellists under Chapter 19 of NAFTA and the Indicative List of Panelists of the World Trade Organization.

51 "Grits OK NAFTA, but No Energy Deal," *Times-Colonist*, December 3, 1993; Ian Austen, "Trade Agreement 'Not Perfect,' but Chrétien Accepts," *Calgary Herald*, December 3, 1993, A1.

52 Cyndee Todgham Cherniak, "Have Senators Clinton and Obama Done Their Homework?," Lang Michener Publications, accessed June 8, 2009, www.langmichener.ca.

53 Jonathan Tasini, "Who Knew NAFTA Had So Many Enemies?," Labor Research Association, February 27, 2008, posted on communicatinglabourrights.wordpress.com/2008/02/28/us-who-knew-nafta-had-so-many-enemies/.

54 Norma Greenaway, "Anti-NAFTA Talk Cause for Concern: Trade Minister," *Canwest News Service*, February 27, 2008; Steven Chase, "Ottawa Plays Oil Card in NAFTA Spat," *Globe and Mail*, February 28, 2008, A1; Susan Delacourt, "PM Issues Warning to NAFTA Foes," *Toronto Star*, April 23, 2008, www.thestar.com.

55 Mike Blanchfield, "Harper Delighted Obama Won't Reopen NAFTA," *Canwest News*, April 22, 2009, accessed June 5, 2009, www.canada.com.

56 Campbell Clark and Rhéal Séguin, "Ottawa Pushes for New Chapter in Free Trade with US," *Globe and Mail*, June 3, 2009, A1.

57 Cyndee Todgham Cherniak, interview by author, Lang Michener office, Toronto, August 21, 2008.

Chapter 5

1 Shawn McCarthy, "'Anti-petroleum' Movement a Growing Security Threat to Canada RCMP Say," *Globe and Mail*, Feb 17, 2015

2 "Shaping Alberta's Future," Executive summary, Alberta Premier's Council for Economic Strategy, May 5, 2011, alberta.ca/premierscouncileconomicstrategy.cfm.

3 "CAPP Crude Oil Forecast," CAPP, June 9, 2014, www.capp.ca.

4 Yadullah Hussain, "Almost $60 billion in Canadian Projects in Peril as 'Collapse' in Oil Investment Echoes the Dark Days of 1999," *Financial Post*, January 2, 2015; Oilsands Review Staff, "IEA Forecasting Canadian Oil Supply Growth of 810,000 BBLS/D to 2020: Delays in New Oilsands Projects," Canadian Oilsands Navigator, Feb 18, 2015, updated April 17, 2015.

5 "No to a New Tar Sands Pipeline," Editorial, *New York Times*, April 3, 2011, WK9, www.nytimes.com.

6 "Statistical Review of World Energy 2009," British Petroleum, 2004 stats (oil only), www.bp.com; Author's calculations.

7 On top of Sands projects, Alberta uses coal and natural gas to produce electricity.

8 "EPA Letter to US State Department," July 16, 2010, 2–3, yosemite.epa.gov.

9 There is a wide range of estimates of the extra GHGs from Sands oil compared to conventional oil. They range from a low of 3% by Ezra Levant, to a high of 200% by Pembina Institute. The EU estimate is 23% higher.

[10] "Carbon Capture and Storage in the Alberta Oil Sands—A Dangerous Myth," WWF-UK, 47.

[11] J. David Hughes, *Will Natural Gas Fuel America in the 21st Century?*, (Santa Rosa, California: Post Carbon Institute, 2011), 50.

[12] Arthur Neslen, "New Hopes That Tar Sands Could be Banned from Europe," *Guardian* (London), February 19, 2015, www.theguardian.com.

[13] "Low Carbon Fuel Standard Program," California Environmental Protection Agency, accessed March 4, 2014, www.arb.ca.gov/fuels/lcfs/lcfs.htm.

[14] Stephen Harper, "Address by the Prime Minister at the Canada –UK Chamber of Commerce," London, July 14, 2006, www.pm.gc.ca/eng/media.asp?category=2&rid=1247.

[15] Anders Hayden, "The UK's Decision to Stop Heathrow Airport Expansion: Sufficiency, Ecological Modernization, and Core Political Imperatives," Draft paper presented at the "Energy, the Environment and New Markets as a Solution" session at the Canadian Political Science Association Meetings at Wilfrid Laurier University, Waterloo, Ontario, May 18, 2011. Used with permission.

[16] Antony Froggatt and Glada Lahn, *Sustainable Energy Security: Strategic Risks and Opportunities for Business*, (London: Chatham House-Lloyd's 360 Risk Insight White Paper, 2010), 9, www.lloyds.com/360.

[17] Richard Heinberg, *Peak Everything: Waking Up to the Century of Declines*, (Gabriola Island, BC: New Society Publishers, 2010), XX-XXI. The next three paragraphs are based on Heinberg's book.

[18] Jeff Rubin, *Why Your World is About to Get a Whole Lot Smaller*, (Toronto: Random House Canada, 2009), 4–5.

[19] Hughes, J. David, *Drill, Baby, Drill: Can Unconventional Fuels Usher in a New Era of Energy Abundance?*, (Santa Rosa: Post Carbon Institute, 2013), 44, fig. 37, http://shalebubble.org/drill-baby-drill/.

[20] "The Dutch Disease," *The Economist*, Nov 26, 1977, 82–3; Paul Krugman, "The Narrow Moving Band, the Dutch Disease, and the Competitive Consequences of Mrs. Thatcher," *Journal of Development Economics* 27, no. 1–2 (1987), 50, doi:10.1016/0304-3878(87)90005-8; Thomas Mulcair, "Tar Sands: Dirty Oil and the Future of a Country," *Policy Options*, March 2012, 62–4.

[21] Ibid.

[22] "Table 282-0088," *CANSIM* (database), Statistics Canada. Using seasonally adjusted numbers from Dec 2002 to Dec 2012, 540,900 jobs were lost in manufacturing.

[23] Nathan Vanderklippe, "Carney Says High Oil Prices Benefit Canada," *Globe & Mail*, September 8, 2012, A13; Kevin Carmichael, "Good Try by Carney, but Traders Like Our Petro-loonie," *Globe & Mail*, April 19, 2012; "Table 282-0088," *CANSIM* (database), Statistics Canada.

[24] T. Clarke, Diana Gibson, Brendan Haley and Jim Stanford, *Bitumen Cliff: Lessons and Challenges of Bitumen Mega-developments for Canada's Economy in an Age of Climate Change*, (Ottawa: Canadian Centre for Policy Alternatives and Polaris Institute, 2013).

[25] Ibid.

[26] Ibid; "Table 282-0088," *CANSIM* (database), Statistics Canada. By 2012, manufacturing jobs had largely recovered in Alberta. The Dec 2012 job level was only 3.2% below Dec 2002 job level; Jason Foster, "Manufacturing Gone Missing," *Our Times*, September 15, 2008, 2.

[27] Ryan Macdonald, "Not Dutch Disease, It's China Syndrome," *Canadian Economic Observer*, 11-010 (2007), www.statcan.gc.ca/pub/11-010-x/00807/10305-eng.htm.

[28] Ibid.

[29] Michel Beine, Charles S. Bos and Serge Coulombe, "Does the Canadian Economy Suffer from Dutch Disease?," *Resource and Energy Economics*, 2012.

[30] Clarke et al, *Bitumen Cliff*, part 2, 1.

[31] Ibid., Table 1. The oil and gas sector has 0.48 jobs per $1 million GDP, while manufacturing has 9.64 jobs. The latter has 20.083 times as many jobs per million GDP; Ibid. Construction generates 7.69 jobs per $1 million GDP; "Table 282-0088," *CANSIM* (database), Statistics Canada.

[32] "Canada's Kyoto Protocol Targets and Obligations," Canada, www.climatechange.gc.ca. Canada agreed to the Kyoto targets in 1996, but did not sign on until 2005.

[33] Will Campbell, "Oil, Gas Sector is Now Canada's Biggest Source of Greenhouse Gases," *Toronto Star*, April 12, 2014.

[34] "Table 282-0088," *CANSIM* (database), Statistics Canada.

[35] "Shaping Alberta's Future," Alberta Premier's Council, 5.

[36] *Sounding an alarm for Alberta*, Policy Options, September 2006, 6.

[37] David J. Climenhaga, "Oh Dear! Beloved Peter Lougheed Calls for Higher Corporate Taxes and Slow Oilsands Development," Alberta Diary, May 12, 2011, www.albertadiary.ca.

[38] *Green Energy Plan*, Alberta's NDP, n.d.

[39] "Special Feature," *Alberta Oil* 4, no. 1 (2007), 35.

[40] Watkins, "The Staples Theory Revisited," *Journal of Canadian Studies*, 12(5) (1977), 83–95. Watkins turned his take on the staples approach to a socialist one in this article.

[41] Tony Clarke et al, *Bitumen Cliff*, 53–4.

[42] Terisa E. Turner and Diana Gibson, *Back to Hewers of Wood and Drawers of Water: Energy, Trade and the Demise of Petrochemicals in Alberta*, (Edmonton: Parkland Institute, 2005), 22.

[43] Helge Ryggvik, *The Norwegian Oil Experience: A Toolbox for Managing Resources?*, (Oslo: Centre for Technology, Innovation and Culture, University of Oslo, 2010), 5.

[44] Henry George, *Progress and Poverty*, 1879 (originally the National Single Tax League Publishers), books.google.ca.

[45] Ryggvik, *Norwegian Oil Experience*, 5.

[46] Ole Gunnar Austvik, *The Norwegian State as Oil and Gas Entrepreneur: The Impact of the EEA Agreement and EU Gas Market Liberalization*, (Saarbrucken, Germany: VDM Verlag Dr. Muller, 2009), 104.

[47] Ibid., 102–3; Ryggvik, *Norwegian Oil Experience*, 34; Ibid. Ryggvik is summarizing here from White paper #25 (1974), 35.

[48] Austvik, *Norwegian State*, 102.

[49] Ole Gunnar Austvik, "What Norway is Doing with Petroleum Rent Collection and Use," article prepared for *Parkland Post*, December 15, 2006, 4.

[50] Leslie Shiell and Colin Busby, *Greater Saving Required: How Alberta Can Achieve Fiscal Sustainability from its Resource Revenues*, (C.D. Howe Institute, May 2008), 2; *Our Fair Share*, 20, 26; "OECD to Canada: Build a Norwegian Oil Fund," Editorial, *Scandinavian Oil-Gas Magazine*, June 13, 2008, www.scandoil.com. The OECD makes similar recommendations; Petoro doesn't operate fields or directly own the licenses. The State's Direct Financial Interests [SDFI] also has major holdings in oil and gas pipelines and oil and gas processing and refining. It also owns 100% of Gassco, an operator of the gas transportation network. Petoro oversees Statoil's production. www.petoro.no; Ole Gunnar Austvik, *Reflections on Permanent Funds: The Norwegian Pension Fund Experience*, (Toronto: Gordon Foundation, 2007), gordonfoundation.ca.

[51] Regan Boychuk, *Profits, Pressure, and Capitulation*, (Edmonton: Parkland Institute, 2010).

[52] *Oil & Gas Security: Emergency Response of IEA countries*, (Norway: IEA, 2011), 5, www.iea.org/papers/security/norway_2011.pdf.

[53] The province owns most of subsurface rights under the soil of farmers.

[54] Clearwater, *Tar Sands: Canada for Sale*, Video documentary, Edmonton, 2008, Chapter 7; "Facts and Statistics," Alberta Energy, 2, www.energy.alberta.ca/OilSands/791.asp; Statistics were supplied by Mineral Rights Unit of Saskatchewan's Ministry of Energy and Resources, September 22, 2010.

[55] Daniel Tencer, "Norway's Oil Fund Heads For $1 Trillion; So Where Is Alberta's Pot Of Gold?," *Huffington Post Business Canada*, January 25, 2014, www.huffingtonpost.ca.

[56] Regan Boychuk, "Misplaced Generosity: Extraordinary Profits in Alberta's Oil and Gas Industry," (Edmonton: Parkland Institute, 2010), 17.

[57] Norges Bank Investment Management, www.nbim.no/en/; "Heritage Fund Information," Alberta Treasury Board, February 24, 2015, www.finance.alberta.ca/business/ahstf/. Calculated in Canadian dollars on April 4, 2015, Norway's oil fund value was 6.98 trillion kroner.

[58] "Country Analysis Brief," International Energy and Analysis, EIA (US), February 25, 2014, www.eia.gov; "Alberta's Energy Industry: An Overview," (Edmonton: Alberta Government, June 2012), 1, 3.

[59] Jeff Rubin and Warren Lovely, "The Vanishing Consequences of a Parity Exchange Rate," *StrategEco*, CIBC World Markets Inc., June 15, 2007, 5.

[60] The Canadian Press, "Canada bound for oil's 'front ranks': CIBC," Business Edge 7, no. 19 (2007), www.businessedge.ca. Rubin chose a wide percentage spread to take into account of "your view of the investment climate in Kazakhstan and Nigeria."

[61] Shiell and Busby, *Greater Saving Required*, 11. Constant dollar value calculated from Statscan CPI.

62 *Our Fair Share: Report of the Alberta Royalty Review Panel,* Edmonton, September 18, 2007.

63 Diana Gibson, *Selling Albertans Short,* (Edmonton: Parkland Institute), 3. If implemented, the *Fair Share* recommendations would have reduced Alberta's projected royalties on oil and gas fall by $3.8 billion by 2016 or a net $2 billion from current revenue levels.

64 Ricardo Acuña, "Royalty Review Not as Radical as Industry Says," *Vue Weekly,* September 26, 2007.

65 *Our Fair Share,* 27–8.

66 David Thompson and Keith Newman, *Private Gain or Public Interest? Reforming Canada's Oil and Gas Industry,* (Ottawa: Canadian Centre for Policy Alternatives, 2009), 11, www.policyalternatives.ca.

67 Andre Plourde, email correspondence with author, Oct 29, 2010. Plourde considers Boychuk's analysis to be an "excess returns" rather than a rent calculation. Corporate taxes are excluded from the calculation of the government's take. Plourde's main concern is that the calculations are based on annual, aggregate data, not on individual projects. These are "flow returns", meaning that "expenditures" are not tied to the "revenues." All this complicates the interpretation of the results.

68 Regan Boychuk, *Profits, Pressure, and Capitulation,* (Edmonton: Parkland Institute, 2010).

69 David Thompson, *Green Jobs: It's Time to Build Alberta's Future,* (Edmonton: Sierra Club, Greenpeace & AFL, 2010), 18; *Wind Turbines in Denmark,* (Copenhagen: Danish Energy Agency, 2009), 8; Thompson, *Green Jobs,* 3.

70 Ibid., 13.

71 Ibid.

72 *Estimated Impact of the American Recovery and Reinvestment Act,* (Washington, D.C: US Congressional Budget Office, 2012), www.cbo.gov; Thompson, *Green Jobs,* 9.

73 Ibid., 18–19.

74 Ibid., 45, 23.

75 *Alberta Budget 2015,* (Edmonton, Alberta Government, 2015), 110, finance.alberta.ca. The budget was announced by the Prentice Conservatives but was not passed.

76 White House, "Administration Announces Nearly $8 Billion in Weatherization Funding and Energy Efficiency Grants," March 12, 2009, www.whitehouse.gov.

77 Homer-Dixon, "Clean Coal? Go Underground, Alberta," *Globe and Mail,* May 4, 2009.

78 BC Hydro is part of a power-sharing consortium that includes electric utilities in Washington, Oregon, Idaho, California and Alberta. Why not change that to a Canadian only power sharing arrangement?

79 Thompson, *Green Jobs,* 37.

80 Ibid., 25.

81 Ibid., 41.

82 John Richards and Larry Pratt, *Prairie Capitalism: Power and Influence in the New West,* (Toronto: McClelland & Stewart, 1979), 242. It was not a crown corporation and shares were sold to Albertans, but it was a provincial government creation. It was later folded into Encana.

83 Peter Lougheed, "A Thirsty Uncle Looks North," *Globe and Mail,* November 11, 2005, A19.

Chapter 6

1 Suncor began as an American company. Its ownership is now diverse, with mainly US and Canadian institutions holding over 60 percent of its shares.

2 Diana Gibson first made this observation.

3 *How Canadians Perceive Gas Price Hikes,* (Leger Marketing), 1–11 (see chap. 1 n. 82).

4 G. Bruce Doern and Glen Toner, *The Politics of Energy,* (Toronto: Methuen, 1985), 107–8.

5 Ibid., 279–80. In 1980, Ontario bought 25% of US-owned Suncor to support Canadianization. Alberta set up the Alberta Energy Company in 1973 to develop oil and gas in Alberta and lessen Canada's dependence on foreign oil. Alberta had a 50% ownership. AECL was merged in 2002 with PanCanadian Energy to become Encana.

6 Daniel Yergin, *The Prize,* (New York: Simon & Schuster, 1991), 203–5. In 1928, Royal Dutch Shell, Anglo-Persian (now BP), France's CFP (forerunner of today's Total) and a consortium of US oil companies signed an agreement to divide up Iraq's oil and work together in the former Ottoman empire (present day Saudi Arabia, Kuwait, oil emirates and Iraq). It was called the 'Red Line' or Stand Still agreement; The group included Standard Oil of New Jersey and Standard Oil Company of New York now ExxonMobil; Standard Oil of Califor-

nia, Gulf Oil and Texaco now Chevron; Royal Dutch Shell; and Anglo-Persian Oil Company. Italian oil-tycoon Enrico Mattei coined the term; That's Carola Hoyos' classification. She missed Total and ENI; "Supermajordämmerung," *Economist*, August 3, 2013. The day of the huge integrated international oil company is drawing to a close."

[7] "Supermajordämmerung," *Economist*, 4.

[8] "Supermajordämmerung," *Economist*, 2.

[9] Ray Walser, "Meeting Energy Challenges in the Western Hemisphere," Heritage Lectures, no. 1079, May 6, 2008.

[10] Ibid.

[11] The Global Oil Industry Supermajordämmerung," *Economist*, Aug 3, 2013, www.economist.com.

[12] Robert L. Hirsch, "Mitigation of maximum world oil production: Shortage scenarios," *Energy Policy* 36 (2008), 886.

[13] Dick Cheney, Speech, London Institute of Petroleum Autumn lunch, 1999, posted at www. petroleum.co.uk/speeches.htm; Kjell Aleklett, "Dick Cheney, Peak Oil and the Final Count Down," Peak Oil, May 12, 2004, www.peakoil.net//Publications/Cheney_PeakOil_FCD.pdf.

[14] Nordine Ait-Laoussine, "Resource Access and Resource Nationalism," *Middle East Economic Survey*, Vol. L1 #44, November 3, 2008.

[15] Paul Stevens, "National Oil companies and international oil companies in the Middle East: Under the Shadow of government and the resource nationalism cycle," *The Journal of World Energy Law & Business*, 1(1) (2008), 5.

[16] Ibid., 6.

[17] Mark J. Perry, "The Oil Shock of the 1930s: Another Factor?,"*Carpe Diem Blog*, November 15, 2008, mjperry.blogspot.com/2008/11/oil-shock-of-1930s.html; "Historical Crude Oil prices," Inflation Data, accessed June 30, 2010, inflationdata.com. The world price dropped in 1938 and was flat in the late 1940s and early 1950s.

[18] Terry Lynn Karl, *The Paradox of Plenty*, (Berkeley: University of California Press, 1997), 246.

[19] Victor Rodriguez-Padilla, "Nationalism and Oil," *Encyclopedia of Energy* 4, (2004), 185.

[20] Leslie E. Grayson, *National Oil Companies*, (Chichester UK: John Wiley & Sons, 1981). Veba in Germany, CFP in France and ENI in Italy were fully or partly state owned.

[21] Igor Osipov, "Towards Regaining Energy Control: Foreign Investment, Kovykta Project, and the Rise of Gazprom," (Edmonton: University of Alberta School of Business, 2006–7), 11.

[22] Rodriguez-Padilla, "Nationalism and Oil," 189; Michael McCaughan, *The Battle of Venezuela*, (New York: Seven Stories Press, 2005), 154–8; Reid W. Click and Robert J Weiner, "Resource Nationalism Meets the Market: Political Risk and the Value of Petroleum Reserves," *Journal of International Business Studies* 41(2010), 783–4.

[23] Irwin Greenstein, "Growing Resource Nationalism Threatens US Consumers," posted on addwealthnow.com; Barack Obama, "From Peril to progress," *The White House Blog*, Jan 26, 2009, www.whitehouse.gov/blog; Ibid.

[24] Rodriguez-Padilla, "Nationalism and Oil," 183, 181.

[25] "EXPROPIACIóN PETROLERA," Trip Atlas, tripatlas.com/Expropiación%20petrolera.

[26] Pierre Terzian, *OPEC: The Inside Story,* (London: Zed Books, 1985).

[27] Ineke Lock, "The United Nations Centre on Transnational Corporations," (Unpublished manuscript, Summer 2002).

[28] Several 'Dependencia' approaches assumed that Western imperialist countries and corporations exploited developing countries and hindered their development. Raul Prebisch and the United Nations Economic Commission on Latin America (ECLA) were early proponents.

[29] Rodriguez-Padilla,"Nationalism and Oil," 185.

[30] Michel Crozier et al, *The Crisis of Democracy: Report on the Governability of Democracies to the Trilateral Commission,* (New York: NYU Press, 1975), 162.

[31] Gordon Laxer, "The Movement that Dare not Speak its Name: The Return of Left Nationalism/ Internationalism," *Alternative: Global, Local, Political* 26, (2001), 1–32.

[32] Robert J. Weiner and Reid W. Click, "Resource Nationalism Meets the Market: Political Risk and the Value of Petroleum Reserves," *Journal of International Business Studies* 41 (June/July 2010), 784, doi:10.1057/jibs.2009.90.

[33] John Dillon, "Bolivia Emulates Norway; Why Doesn't Canada?," Kairos Briefing Paper, no. 4 (Oct 2006), www.kairoscanada.org/fileadmin/fe/files/PDF/Publications/policyBriefing4 Bolivia0610.pdf; Chronicle editors, "Latin America's Nationalist Mistakes," *Latin Business Chronicle*, www.latinbusinesschronicle.com; "Bolivia Completes Oil and Gas Nationalization," *Mercopress*, Oct 30, 2006, accessed June 21, 2010, en.mercopress.com/2006/10/30/

bolivia-completes-oil-and-gas-nationalization.
[34] Ibid., 19–20.
[35] Pete Stark, "The Winds of Change: Resource Nationalism Shifts the Balance of Power to National Oil Companies," *JPT Online* 59, no. 1 (January 2007), www.spe.org/spe-app/spe/jpt/2007/01/guest_editorial .htm.
[36] Wong, "Government Ownership: Why this Time It Should Work," *The McKinsey Quarterly,* June 2009, accessed June 26, 2009, www.mckinseyquarterly.com.
[37] CommDev, "Oil Industry's Increasing Focus on CSR," March 4, 2010, commdev.org/content/article/detail/2590/; David Lertzman et al, "A National Oil Company as Social Development Agent," *International Review of Business Research Papers* 5, no. 5 (2009), 6; "Statoil," Academic Dictionaries and Encyclopedias, en.academic.ru/dic.nsf/enwiki/116601.
[38] Ivonne Yanez, intervewed by Amy Goodman, "Keep the Oil in the Soil: Ecuador Seeks Money to Keep Untapped Oil Resources Underground," at the Copenhagen Climate Summit 2009, full transcript and 10 min. video, posted on Democracy Now, December 11, 2009, www.democracynow.org; Hintadupfing, "The Ecuadorian Government Signed a Landmark deal to Prevent Drilling," August 5, 2010, hintadupfing.blogspot.com/2010/08/ecuador-signs -historic-deal-to-leave.html; "Ecuador: Yasuni Campaign Remains under Threat Despite Progress," Eye on Latin America, March 21, 2014, eyeonlatinamerica.wordpress com/2014/03/21/yasuni-threat-despite-progress/.
[39] Ineke Lock, "A Comparative Study of Approaches to Corporate Social Responsibility," (PhD thesis, University of Alberta, Sociology 2010), chap. 1.
[40] Michael Klare, "The Oil Catastrophe," *Nation,* May 28, 2010, www.thenation.com.
[41] Milton Friedman, "The Social Responsibility of Business is to Increase its Profits," *New York Times Magazine,* September 13, 1970.
[42] David Thompson and Keith Newman, *Private Gain or Public Interest: Reforming Canada's Oil and Gas Industry,* (Edmonton: Parkland Institute and CCPA, 2009), 13.
[43] George Monbiot, "The Climate Denial Industry Is Out to Dupe the Public. And It's Working," *Guardian* (London), December 7, 2009, www.theguardian.com; Donald Gutstein, "This Is How You Fuel a Community of Climate Deniers," *The Tyee,* December 10, 2009, thetyee.ca/Books/2009/12/10/Climate Deniers/; Donald Gutstein, *Not a Conspiracy Theory: How Business Propaganda Hijacks Democracy,* (Toronto: Key Porter Books, 2009); Exxon became ExxonMobil in 1999; Gutstein, "Fuel a Community," 6.
[44] Charles McPherson and Stephen MacEarraigh, "Corruption in the Petroleum Sector," World Bank, 2007; "Horacio Marquez Says Suncor (SU) Will Rebound Strongly," Green Stocks Central, July 23, 2008, greenstockscentral.com/page/304; "Information Letter 2009-37," Alberta Energy, p. 3, inform.energy.gov.ab.ca/Documents/Published/IL-2009-37.pdf; "Table 5.5f," *Statistical Handbook,* CAPP, www.capp.ca/getdoc.aspx?DocId=167463&DT=NTV. The par prices in the Alberta Energy letter were converted from cubic metres to barrels. The price differentials between sweet light crude and extra heavy oil are from "Table 5.5f"; *Existing and Proposed Canadian Commercial Oil Sands Projects,* (Calgary: Strategy West, 2009), www.strategywest.com. At $16.50 a barrel, Suncor's superprofits would be about $2.65 billion a year.
[45] "Tables 4.2.1., 6.3 and 6.4," *Energy Statistics Handbook,* StatsCanada, Catalogue #57-601-X, first quarter 2012, Percentages derived from the aforementioned tables. BC produces 22 percent of Canada's natural gas.
[46] John Michael Greer, *The Long Descent,* (Gabriola Island: New Society Publishers. 2008), 13.
[47] Danny Williams, "Newfoundland Won't Be Held Ransom by Big Oil Companies," *Globe and Mail,* April 5, 2010, A15.
[48] Peter Sinclair, "An Ill-wind Is Blowing Some Good," Memorial University, February 2008, published as part of the "Oil, Power and Dependency" project; Erin Weir, "Danny Williams and Oil Royalties," *The Progressive Economics Forum,* Aug 25, 2007, www.progressive-economics.ca/2007/08/25/williams-oil-royalties/; "Global Energy Company Rankings," Platts Top 250, 2014, www.platts.com.
[49] Thompson and Newman, *Private Gain or Public Interest,* 22.
[50] PetroStrategies, 'Leading Oil and Gas Companies Around the World'. 2007 Ranked by oil equivalent reserves. http://www.petrostrategies.org/Links/Worlds_Largest_Oil_and_Gas_Companies_Sites.htm Cit., 16.
[51] Thompson and Newman, *Private Gain or Public Interest,* 20.

[52] Ibid., 17.

[53] "Federal Government Quietly Releases $490B Military Plan," CBC, www.cbc.ca.

[54] Keith Newman, correspondence with author, July 26, 2010.

[55] Mel McMillan, correspondence with author, July 23, 2010.

[56] Healy, Paul M. and Krishna G. Palepu, "The Fall of Enron," *Journal of Economic Perspectives* 17, no. 2 (Spring 2003), 3–26.

[57] Kevin Rudd returned to power briefly in 2013 after ousting Prime Minister Julia Gillard, his successor. He lost the next election.

[58] *The BC Energy Plan*, Government of British Columbia, April 9, 2009, www.energyplan.gov.bc.ca.

[59] "One Day," City of Vancouver, vancouver.ca/oneday/takeAction/atHome/incentive.htm.

[60] Jeff Bell and Tim Weis, *Greening the Grid: Powering Alberta's Future with Renewable Energy*, (Drayton Valley, Alberta: Pembina Institute, January 2009).

[61] "Utilities," *The Canadian Encyclopedia*, www.thecanadianencyclopedia.com.

Chapter 7

[1] *Energy East—Backgrounder*, Environmental Defence, March 18, 2014, environmentaldefence.ca

[2] William Kilbourn, *Pipeline: TransCanada and the Great Debate*, (Toronto: Clarke, Irwin & Company, 1970), vii.

[3] Ibid., 24–5.

[4] Petr Cizek, *West-to-East Natural Gas Pipelines circa 1970* [map], Cizek Environmental Services, 2014, www.cizek.ca, data source: The National Atlas of Canada.

[5] Petr Cizek, *Current Natural Gas Pipelines to the US* [map], Cizek Environmental Services, 2014, www.cizek.ca, data source: The National Atlas of Canada.

[6] "Table 6.1," *Energy Statistics Handbook*, Statistics Canada, fourth quarter 2011. Calculated by John Dillon, the exact percentage is 64.2%.

[7] Nova Scotia, New Brunswick, Newfoundland and PEI all lack province-wide natural gas distribution systems. The Maritimes & Northeast Pipeline (M&NP) exports gas at St. Stephen, NB; Angela Carter, email correspondence with author, Aug 6, 2012. Carter is a Political Science professor at the University of Waterloo; Nathan Vanderklippe, "A Pipeline in Peril," *Globe & Mail*, July 10, 2011, www.theglobeandmail.com.

[8] Tanya Whyte, *The Political Economy of Canadian Oil Export Policy, 1949–2002*, (Political Science MA thesis, University of Alberta, 2010), 17. This was after the route south of the great lakes had been built.

[9] Petr Cizek, *Canadian Oil Pipelines circa 1977* [map], Cizek Environmental Services, 2014, www.cizek.ca, data source: The National Atlas of Canada.

[10] "2009 Proposed Pipeline," Oil Sands Truth: Shut Down the Tar Sands, accessed July 20, 2011, oilsandstruth.org/2009-proposed-pipelines.

[11] "Details on How Enbridge Will Expand Capacity of Alberta Clipper Oil Sands Crude Pipeline without US Review." Green Car Congress, August 23, 2014, www.greencarcongress.com/2014/08/20140823-enbridge.html.

[12] Petr Cizek, *Current and Proposed Oil Pipelines from Canada to the US* [map], Cizek Environmental Services, 2014, www.cizek.ca, data source: The National Atlas of Canada.

[13] Ontario Ministry of Energy, correspondence with author, April 17, 2015; www.enbridge.com. The combined Canadian and US oil now meets 82 percent of oil supplies to Ontario refiners.

[14] James Laxer, *Oil and Gas*, (Toronto: Lorimer, 1983), 8.

[15] John N. McDougall, *Fuels and the National Policy*, (Toronto: Butterworths, 1982), 59–64.

[16] Imports were 136,729 cubic metres per day in 1973. By 1983, these had fallen to 39,322 cubic metres per day. "Oil Imports 1947-1985," *Statistical Handbook 2014*, CAPP, 195.

[17] Larry Hughes, "Eastern Canadian Crude Oil Supply and Its Implications for Regional Energy Security," *Energy Policy* (2010), 3–5, doi:10.1016/j.enpol.2010.01.015.

[18] The world's first modern oil well was discovered in Petrolia, Canada West, in 1858 in colonial Canada.

[19] A small portion of the Bakken field is in Saskatchewan.

[20] Justine Hunter and Brent Jiang, "BC Lowers Expectations for LNG Windfall," *Globe and Mail*, October 22, 2014, A1.

[21] Petr Cizek, *Proposed Offshore Natural Gas Pipelines* [map], Cizek Environmental Services, 2014, www.cizek.ca, data source: The National Atlas of Canada.

[22] Petr Cizek, *Proposed Offshore Oil Export Pipelines* [map], Cizek Environmental Services, 2014, www.cizek.ca, data source: The National Atlas of Canada.

[23] Nathan Vanderklippe and Shawn McCarthy, "TransCanada Eyes Pipeline to East Coast," *Globe*, March 23, 2012, B1.

[24] The line is to begin at Bruderheim, 60 km northeast of Edmonton; Samantha Garvey, "Slick Talk: A History of Oil Tankers and BC's Coast," *Terrace Standard*, August 22, 2012; Alexis Stoymenoff, "BC Pipeline Opposition Joined by Eastern Counterparts in Ottawa," *Vancouver Observer*, March 13, 2012, www.vancouverobserver.com.

[25] Matt Price and Gillian McEachern, *Freedom from Dirty Oil: Ontario's Tar Sands Decision*, (Toronto: Environmental Defence and Forest Ethics, January 2009), 16, forestethics.org/downloads/pipeline-web.pdf.

[26] That proposal has since been shelved.

[27] Some Venezuelan and Mexican oil is as heavy and carbon laden as tar sands oil.

[28] Yadullah Hussain, "Pipeline Plan to Send Crude from Montreal to Maine Raised Ire in New England," *Financial Post*, May 22, 2013.

[29] Tribalscribal and Shelley Kath, "Approval of Line 9B Bringing Tar Sands Oil to New England's Doorstep," Indy Media Climate, March 10, 2014.

[30] Carrie Tait, "Enbridge Assailed over Spill Response," *Globe*, July 11, 2012, B1.

[31] Michael Schwirtz, "Oil Pipeline Ruptures in Arkansas," *New York Times*, March 30, 2013, www.nytimes.com.

[32] Erin Flegg, "Changes to Navigable Waters Protection Act Dangerously Undermine Environmental Protection, Say Critics," *Vancouver Observer*, January 1, 2013, www.vancouverobserver.com.

[33] Tim Groves, "Idle No More Events in 2012: Events Spreading across Canada and the World," The Media Co-op, December 27, 2012, www.mediacoop.ca/story/idle-no-more-map-events-spreading-across-canada-an/15320.

[34] Justin Giovannetti, "Lac-Mégantic Reacts to Transportation Safety Board Report," *Globe and Mail*, August 22, 2014.

[35] Andy Blatchford, "Why Was Lac-Mégantic Crude Oil So Flammable? Authorities Want Closer Look at Cargo from Train Disaster," *National Post*, August 2, 2013.

[36] Claudia Cattaneo, "TransCanada Mulls Switching Natural Gas Mainline to Oil Service," *Financial Post*, April 27, 2012, Insight West Research www.insightwest.ca/news/gas-insight/transcanada-mulls-switching-natural-gas-mainline-to-oil-service.

[37] Canadian Press, "National Energy Board Approves New Tolls for TransCanada's Mainline Pipeline," *Financial Post*, March 27, 2013, business.financialpost.com.

[38] Shawn McCarthy and Jeffrey Jones, "The Promise and the Perils of a Pipe to Saint John," *Globe and Mail*, August 2, 2013, B1.

[39] "Irving Oil and TransCanada Announce Joint Venture to Develop New Saint John Marine Terminal," News Release, Irving Oil, August 1, 2013, www.irvingoil.com. TCL is partnering with Irving oil to develop a deep water port in Saint John to enable Canadian oil exports via the world's largest oil tankers.

[40] Michael Tutton, "Prime Minister Says Pipeline to East Coast Helps Build Canadian Energy Security," *Maclean's*, August 8, 2013.

[41] Shawn McCarthy and Jeffrey Jones, "The Promise and Perils of a Pipe to Saint John," *Globe and Mail*, August 2, 2012, B1.

[42] Kevin Bassett, "Atlantic Canada Needs to Reduce Oil Use, not Build Pipeline: Professor," *Calgary Herald*, July 9, 2013. Larry Hughes argues this. This may be generally true. But if ocean spills are less frequent, they are often more devastating than land spills.

[43] Richard Gilbert, "Why Quebec is Poised for an Oil Shock," *Globe and Mail*, September 30, 2010, www.theglobeandmail.com.

[44] Patrick Bonin, interview by author, Montreal, June 29, 2011.

[45] Gary Park, "Looks at Converting Gas Line to take Bakken Oil to Refineries Paying Top Prices,"*Petroleum News* 17, no. 19, Week of May 6, 2012, www.petroleumnews.com/pntruncate/303075326.shtml.

[46] J. David Hughes, "A Reality Check on the Shale Revolution," *Nature* 494, February 21, 2013, 307–8.

[47] Ross Marowits, "Gaz Metro Leads Opposition to TransCanada's Energy East Project," *National Post*, October 16, 2014.

[48] "No Tar Sands Pipeline through Northern Ontario," Change.org, accessed Aug 3, 2013, www.change.org/en-CA/petitions/no-tar-sands-pipeline-through-northern-ontario-3.

[49] Quebec got 54% of Canada's oil imports. Based on Larry Hughes' calculations of available

refined petroleum products in the three regions. Hughes calculates a net figure of imported petroleum used in each region after factoring in exports and imports of refined petroleum. See Table 6 in Hughes, "Eastern Canadian crude," p. 5. Availability numbers were multiplied by percentage of oil imports in each region. Ontario 20%, Quebec 88%, Atlantic 83%. StatsCan, April 2011.

[50] Louis-Gilles Francoeur, "Le Québec pourrait manquer de pétrole," Le Devoir, February 5, 2008, 1.

[51] Louis-Gilles Francoeur, "Sécurité énergétique," Le Devoir, February 7, 2008, 4.

[52] Louis-Gilles Francoeur, "Sécurité énergétique: Rabaska affaiblirait le Québec," Le Devoir, February 8, 2008, 1.

[53] Ibid.

[54] David Nesseth, "Bill 37 Tabled in Quebec to Ban St. Lawrence Fracking Ahead of Study," HazMat, May 22, 2013, www.hazmatmag.com.

[55] Monique Beaudoin, "Quebec government puts brakes on shale-gas drilling," Postmedia News, March 29 2011. Not all petition supporters were from Quebec's south shore.

[56] Lucie Sauvé, interview by author, June 28, 2011. Sauvé is on the environmental committee of Saint-Marc sur Richelieu, where she lives. The town is where opposition to gaz de schiste began.

[57] Justin Brake and Daniel Miller, "Fracking Moratoria could Cause "Domino Effect" in North America," The Independent (Nfld), December 20, 2014.

[58] Larry Hughes, "Nova Scotia's Energy Strategy: Energy Security for Nova Scotians," April 2, 2009, 13, available at http://lh.ece.dal.ca/enen.

[59] Melanie Martin, "Churchill Falls: The 1969 Contract," 2006, www.heritage.nf.ca/articles/politics/churchill-falls.php.

[60] Lynn Moore, "Newfoundland Challenges Churchill Falls Hydro Deal with Quebec," Montreal Gazette, November 30, 2009.

[61] "Refinery shutdown a warning to be better prepared for future shortages," Amherst Daily News, ns.dailybusinessbuzz.ca/Provincial-News/2011-08-02/article-2687069/NS%3A-Gas-shortage-sends-message/1.

[62] Larry Hughes, "Eastern Canadian Crude"; "Table 6," Total oil production, Barrels, Newfoundland & Labrador Statistics Agency. Industry Statistics oil production. Total Oil Production, Barrels Newfoundland and Labrador November 1997 to Date, www.stats.gov.nl.ca/Statistics/Industry/. Output was 368,443 bpd in 2007.

[63] Larry Hughes, "Energy Security and Crude Oil in Atlantic Canada," February 6, 2012, dclh.electricalandcomputerengineering.dal.ca/enen/index.html.

[64] "Come By Chance Refinery Now Processing Oil Pumped Off Newfoundland," CBC, May 20, 2015.

[65] Some Newfoundland oil reaches Montreal via the Portland Maine to Montréal pipeline. Without access to refiners' books, it's impossible to know how much Newfoundland oil was consumed in Canada.

[66] Hughes, "Eastern Canadian Crude," 4.

[67] Cape Breton Post, "ExxonMobil Says Not Economical to Expand Sable Island Gas Project," Cape Breton Post, July 9, 2010, www.capebretonpost.com.

[68] "Table 6.3," Energy Statistics Handbook 2010, Statistics Canada, www.statcan.gc.ca; "Canada's Population Clock," Statistics Canada. On Aug 4, 2011, Canada's estimated population was 34,539,151 and Atlantic Canada's was 2,355,694; Report of the Canada Power Grid Task Force, Vol II, Canadian Academy of Engineering, 13 and Table 1-1. Available at www.acad-eng-gen.ca.Newfoundland produced 40,048 GWh of hydro-electricity in 2007, while all of the Atlantic provinces consumed 42,000 GWh the same year. NB and NS produced an additional 3,728 GWh. See p 13 and Table 1-1. Available at www.acad-eng-gen.ca; Larry Hughes, "Nova Scotia's Energy Strategy: Energy Security for Nova Scotians," April 2, 2009, 13, www.lh.ece.dal.ca/enen. Muskrat Falls, the first of the two Lower Churchill Falls sites to be developed, will produce five terawatt hours of electricity a year, an addition of about 12% to Newfoundland's total.

[69] "Green" Hydro Power Understanding Impacts, Approvals, and Sustainability of Run-of-River Independent Power Projects in British Columbia," Tanis Douglas Watershed Watch Salmon Society, August 2007, www.watershed-watch.org/publications/files/Run-of-River-long.pdf.

[70] Canada Power Grid Task Force, Academy of Engineering. Calculated from Table 1-2 installed generating capacity in each province in 2007.

[71] "Muskrat Falls Could Alter Power Plans," Ocean Resources Online, June 15, 2011, www.ocean-resources.com/industry-news.asp?newsid=10027.

72 *Cable Routes from Labrador and Newfoundland to Atlantic Maritime Provinces* [map], data source: www.nalcorenergy.com.

Chapter 8

1 Canadian scholars, "Acting on Climate Change: Solutions from Canadian Scholars," Dialogues on Sustainability, www.sustainablecanadadialogues.ca/en/scd.
2 Richard Fidler, "Campaign Against Alberta Tar Sands Given Massive Boost by Quebec Public," *Canadian Dimension*, November 28, 2014.
3 Martin Lukacs, "Indigenous Rights Are the Best Defence Against Canada's Resource Rush," *Guardian (London)*, April 26, 2013, www.theguardian.com.
4 Bertrand Russell, *In Praise of Idleness: And Other Essays*, (London: Routledge, 2004), 11.
5 They include Thomas Princen, Robert & Edward Skidelsky, Juliet Schor, Wolfgang Sachs and Anders Hayden.
6 Thomas Princen, *The Logic of Sufficiency*, (Cambridge, MA: MIT Press, 2005), 18.
7 Robert and Edward Skidelsky, *How Much is Enough? The Love of Money, and the Case for the Good Life*, (London: Penguin Books, 2012), 80.
8 Edmund Burke, "Speech on Independence of Parliament," 1780.
9 Edmund Burke, *A Vindication of Natural Society: A View of the Miseries and Evils Arising to Mankind*, 1756.
10 John Maynard Keynes, *Essays in Persuasion*, (London: Macmillan & Co, 1931). The article was based on Keynes' 1928 lecture, gutenberg.ca/ebooks/keynes-essaysinpersuasion/keynes-essaysinpersuasion-00-h.html.
11 Robert and Edward Skidelsky, *How Much is Enough? The Love of Money, and the Case for the Good Life*. (London: Penguin Books, 2012), 54.
12 Ibid., 69, 12.
13 Ibid., 89–90.
14 Thomas Princen, *The Logic of Sufficiency*, (Cambridge, MA: MIT Press, 2005), xii, 13.
15 Jennifer Karns Alexander, *The Mantra of Efficiency from Waterwheel to Social Control*, (Baltimore: John Hopkins University Press, 2008), 65–8.
16 Princen, *Logic of Sufficiency* , xi–xii.
17 David Rosnick, "Reduced Work Hours as a Means of Slowing Climate Change," (Washington DC: Center for Economic and Policy Research, Feb 2013).
18 Juliet Schor, "Less Work, More Living: Working Fewer Hours Could Save our Economy, Save Our Sanity, and Help Save Our Planet," *Yes Magazine*, September 2, 2011.
19 George Monbiot, *Heat: How to Stop the Planet Burning*, (London: Penguin Books, 2006), 215.
20 Jeff McMahon, "How Japan Discovered Conservation: Fukushima," *Forbes*, January 6, 2013, www.forbes.com.
21 Graham Turner, "A Comparison of The Limits to Growth with Thirty Years of Reality," (*Socio-Economics and the Environment in Discussion*, CSIRO working paper, June 2008), 1–2.
22 Anders Hayden, *When Green Growth is Not Enough: Climate Change, Ecological Modernization, and Sufficiency*, (Montreal: McGill-Queen's University Press, 2014), 123.
23 Herman Daly, "A Steady-State Economy," Sustainable Development Commission (UK), *The Oil Drum*, May 5, 2008, www.theoildrum.com/node/3941.
24 Herman E. Daly, "Economics in a Full World," *Scientific American Special Issue*, September 2005.
25 Hayden, *Green Growth*, 319.
26 Nikiforuk, *The Energy of Slaves*, 221.
27 Stephen R. Tully, "The Contribution of Human Rights to Universal Energy Access," *Northwestern University Journal of International Human Rights* 4, no. 3 (2006).
28 *World Energy Assessment: Energy and the Challenge of Sustainability*, UN Development Programme, 2000.
29 Tully, "Contribution of Human Rights."
30 Energy Planning: A Guide for Northwest Indian Tribes, Northwest Sustainable Energy for Economic Development, n.d., p. 6, fig.1-1, www.nwseed.org/wp-content/uploads/2013/05/NWSEED_Tribal-GB_Final.pdf.
31 Ibid., paragraph 62; "The World's Biggest Public Companies," *Forbes*, www.forbes.com/global2000/list/.
32 Brenda Boardman, *Fixing Fuel Poverty*, (London: Earthscan, 2010).
33 Monbiot, *Heat*, 65.
34 *Statistical Bulletin: Excess Winter Mortality in England and Wales, 2011/12*, UK Office for National

Statistics, www.ons.gov.uk.

35 *The Wrong Direction: How UK Fuel Poverty Policy Lost Its Way*, National Energy Action (UK), May 2008, 21, www.nea.org.uk.

36 David Korten, *The Great Turning: From Empire to Earth Community*, (Bloomfield, Conn: Kumerian Press, 2006), 32.

37 Ibid. Adapted from Korten, *The Great Turning*.

38 Jeremy Rifkin, *The Third Industrial Revolution*, (New York: Palgrave Macmillan, 2011), 14.

39 Keynes, "National Self-Sufficiency", *The Yale Review* 22, no. 4 (June 1933), 755–769.

40 Ibid. The following quotes and excerpts are taken from this.

41 Examples are The Eco-nationalist Platform in Ireland and Australians Against Further Immigration or AAFI party, which declared itself eco-nationalist.

42 Arran E.Gare, *Postmodernism and the Environmental Crisis*, (London: Routledge, 1995). 143–152.

43 Anders Hayden, *When Green Growth is Not Enough: Climate Change, Ecological Modernization, and Sufficiency*, (Montreal: McGill Queen's University Press, 2014), chapter 11.

44 Herman Daly, "Farewell Speech," January 14, 1994, www.whirledbank.org/ourwords/daly.html.

45 Hayden, *Green Growth*, chapter 3. Much of the next few paragraphs are based on this chapter.

46 Ibid., 83.

47 Peter Victor, "Growth, Degrowth and Climate Change: A Scenario Analysis," *Ecological Economics* 84 (2012), 206–212.

48 Quoted in Hayden, *Green Growth*, 295.

49 Victor, "Growth, Degrowth," 209–10.

50 Mathis Wackernagel and William Rees, *Our Ecological Footprint: Reducing Human Impact on the Earth* (Vancouver: New Society Publishers, 1996).

51 Thomas Homer-Dixon and Nick Garrison, "Introduction," in *Carbon Shift: How the Twin Crises of Oil Depletion and Climate Change Will Define the Future*, ed. Homer-Dixon, (Toronto: Random-House Canada, 2009), 4–5.

52 Thomas Homer-Dixon, "The Tar Sands Disaster," *New York Times*, March 31, 2013.

53 Hayden, *Green Growth*, 87.

54 Janet Eaton, "What is Degrowth?," Beyond Collapse Blog, August 6, 2012, beyondcollapse. wordpress.com.

55 Ian Angus, "How to Make an Ecosocialist Revolution," Presentation to the Climate Change Social Change conference, Melbourne, Australia, October 2, 2011.

56 Clayton Thomas-Muller, "The Rise of the Native Rights-Based Strategic Framework," *Canadian Dimension Magazine* 47, no. 3 (2013).

57 Hayden, *Green Growth*, 100.

58 Anders Hayden, "Enough of That Already: Sufficiency-Based Challenges to High-Carbon Consumption in Canada," (unpublished paper, 2013).

59 Hayden, *Green Growth*, 131.

60 Ibid., 153.

Chapter 9

1 Jeff Goodell, "Climate Change and the End of Australia," *Rolling Stone*, October 3, 2011, www.rollingstone.com.

2 "Unburnable Carbon—Are the World's Financial Markets Carrying a Carbon Bubble?," Carbon Tracker Initiative, London, Mar 2012, www.carbontracker.org. Scientists calculate that if emissions are limited to 565 more GTt CO2 between 2011 and 2050 there is a 20% chance global warming will exceed 2 degrees Celsius; *WEO 2012*, (Paris: IEA, 2012), 25. The IEA warns that "no more than one-third of proven reserves of fossil fuels can be consumed prior to 2050 if the world is to achieve the 2 degrees C goal"; Bill McKibben, "Global Warming's Terrifying New Math," *Rolling Stone*, August 2, 2012; Jeff Rubin, *The End of Growth: But Is that All Bad?*, (Toronto: Random House Canada, 2012).

3 Jeffrey Sachs, "Need Versus Greed," Project Syndicate, February 2, 2013, www.project-syndicate.org/print/need-versus-greed.

4 McKibben, "Terrifying New Math."

5 Ibid.

6 Maarten A. Hajer, *The Politics of Environmental Discourse: Ecological Modernization and the Policy Process*, (Oxford Scholarship Online, 1997), 3, www.oxfordscholarship.com.

[7] Anders Hayden, *Green Growth* (see chap. 8 n. 22).

[8] Patrick Bond, Rehana Dada and Graham Erion, *Climate Change, Carbon Trading and Civil Society,* (University of KwaZulu Press, 2009).

[9] Naomi Klein, *This Changes Everything,* (Toronto: Knopf Canada, 2014), 220.

[10] For example, see Richard Heinberg, *Powerdown: Options and Actions for a Post-carbon World,* (Gabriola Island BC: New Society Publishers, 2004); John Michael Greer, *The Long Descent,* (Gabriola Island BC: New Society Publishers, 2004).

[11] Lawrence D Frank et al, "Many Pathways from Land Use to Health," *Journal of the American Planning Association* 72, no. 1 (Winter 2006), 75–87.

[12] Richard Gilbert and Anthony Perl, *Transport Revolutions: Moving People and Freight without Oil,* (London: Earthscan, 2010).

[13] See Mark Jaccard, "Peak Oil and Market Feedbacks: Chicken Little versus Dr. Pangloss," in *Carbon Shift,* ed. Thomas Homer-Dixon, (Toronto: Random House Canada, 2009), 97–134.

[14] William Stanley Jevons, "VII," *The Coal Question* (2nd ed.), (London: Macmillan and Co, 1866).

[15] "Radical New Agenda Needed to Achieve Climate Justice," Global Justice Ecological Project, November 24, 2011, climate-connections.org.

[16] "Background for COP 8," Centre for Science and Environment, October 25, 2002.

[17] Thomas S. Ahlbrandt and Peter J. McCabe, "Global Petroleum Resources: A View to the Future," *Geotimes,* Nov 2002, www.geotimes.org/nov02/feature_oil.html.

[18] "Development of Renewable Energy Sources in Germany 2011," Federal Ministry for the Environment and Nature Conservation and Nuclear Safety (Germany), July 2012, www.irena.org; James Hansen, *Storms of My Grandchildren,* (London: Bloomsbury Press, 2009), 218, 221.

[19] *Efficiency in Electricity Generation,* Eurelectric, July 2003, p. 13, www.eurelectric.org.

[20] Anastassia M. Makarieva, Victor G. Gorshkov and Bai-Lian Li, "Energy Budget of the Biosphere and Civilization: Rethinking Environmental Security of Global Renewable and Non-renewable Resources," *Ecological Complexity* 5, Aug 3, 2008, 281–8, www.bioticregula tion.ru/common/pdf/energy08.pdf.

[21] Andrew Nikiforuk, *The Energy of Slaves: Oil and the New Servitude,* (Vancouver: Greystone Books, 2012), 225.

[22] Rubin, *Your World is About to Get a Whole Lot Smaller.*

[23] Rubin, *End of Growth,* 37. But, China's and India's economies have grown through oil price spikes.

[24] Ibid., 33.

[25] Armine Yalnizyan, *The Rise of Canada's Richest 1%,* (Ottawa: Canadian Centre for Policy Alternatives, 2010), 3.

[26] David Fleming and Shaun Chamberlin, *TEQs: Tradable Energy Quotas,* Parliamentary Report, UK House of Commons All Party Parliamentary Group on Peak Oil, (London: The Lean Economy Connection, 2011), 37, www.teqs.net.

[27] Jeffrey J. Brown, "Is a Net Oil Export Hurricane Hitting the US Gulf Coast?," *The Oil Drum,* June 2, 2008, www.theoildrum.com/node/4092.

[28] David Fleming, *Energy and the Common Purpose,* (London: The Lean Economy Connection, 3rd edition, 2007), 8.

[29] Ibid., 38; Ibid., 39; Aubrey Meyer, *Contraction and Convergence: The Global Solution to Climate Change,* (Totnes, Devon: Green Books, 2005).

[30] Colin Campbell, "The Rimini Protocol, an Oil Depletion Protocol: Heading Off Economic Chaos and Political Conflict during the Second Half of the Age of Oil," *Energy Policy* 34, no. 12 (2006), 1319–25.

[31] Fleming and Chamberlin, *TEQs,*

[32] Ibid., 36.

[33] Rob Hopkins, *The Transition Town Handbook,* (White River Crossing, Vermont: Chelsea Green Publishing, 2008), 76.

[34] "FAQs on TEQs," UK Parliamentary Group on Peak Oil & The Lean Economy Connection, London, 2011, www.teqs.net. Because of their wide-ranging impacts on Britain's economy, further legislation would almost certainly be used to implement TEQs.

[35] Fleming and Chamberlin, *TEQs,* 22.

[36] Frances Moore Lappé, *EcoMind,* (New York, Nation Books, 2011), 83, 99.

[37] "TEQs in summary," The Fleming Policy Centre with the support of the All Party Parliamentary Group on Peak Oil, www.teqs.net/summary/. The next two paragraphs are based on this.

[38] Fleming, *Energy and the Common Purpose*, 12.

[39] Ibid., 14.

[40] Ibid., 20.

[41] David Strahan, *The Last Oil Shock: A Survival Guide to the Imminent Extinction of Petroleum Man*, (London: John Murray, 2007), 230–1.

[42] Units of energy per unit of GDP.

[43] Fleming, *Energy and the Common Purpose*.

[44] Fleming and Chamberlin, *TEQs*, 19–20.

[45] Strahan, *The Last Oil Shock*, 228.

[46] Ibid., 231.

[47] Nell Boase, "Carbon Credits Are 'Wrong' Says Benn," *Guardian* (London), May 31, 2007, www.theguardian.com.

[48] George Orwell, "Shopkeepers At War," essay, 1941, www.orwell.ru/library/essays/lion/english/e_saw.

[49] Ma'anit, personal correspondence with author, July 8, 2013. I am greatly indebted to Ma'anit for raising many of these questions.

[50] *Report of the Canada Power Grid Task Force*, Vol II, (Canadian Academy of Engineering, 2011), 22, Table 1-1.

[51] Fleming and Chamberlin, *TEQs*, 37.

Chapter 10

[1] Work hours are much higher in the US than in Western Europe.

[2] Leo Panitch, "Why are Canada's Trains Vulnerable? Good Old Capitalist Cost-Cutting," *The Bullet*, July 16, 2013, www.socialistproject.ca/bullet/852.php#continue.

[3] Ed Crooks, "US Oil Groups Seek Easing of Drilling Curbs," *Financial Times*, January 4, 2011, www.ft.com.

[4] Stephen Rodrigues, email to author, July 30, 2014, 2013 statistics. Rodrigues is the Manager of Technical / Market Research at CAPP.

[5] *Oil & Gas Security: Canada 2010*, (Paris: IEA, 2010), 5–6. Conventional oil output, including offshore oil, was 1.8 m b/d while natural gas liquids, from which gasoline can be made, was at 700,000 b/d. Stats Canada puts Canada's annual conventional oil output at 1.39 million b/d, while CAPP puts it at 1.84 m b/d. Canadians use 1.7 to 1.8 million barrels a day.

[6] Will Koop, "BC Tap Water Alliance Final Argument, July 14, 2011," (Submission to National Energy Board, Hearing Order GH-1-2011), 2, www.bctwa.org/FrkBC-NEB-FinalArgument-July14-2011.pdf.

[7] "EIA projects strong crude oil, natural gas production through 2035," *Oil & Gas Journal*, January 24, 2012. The EIA (Energy Information Administration) projects that the US will become a net exporter of LNG in 2016.

[8] Thomas Homer-Dixon, *The Upside of Down*, (Toronto: Vintage Canada, 2007), 83.

[9] At projected rates of output.

[10] "Figure 13," *Canadian Energy Overview 2013 - Energy Briefing Note*, National Energy Board, last modified February 4, 2015, www.neb-one.gc.ca/nrg/ntgrtd/mrkt/vrvw/2013/index-eng.html. Used with permission.

[11] "Alberta's Oil Sands," Alberta Government, n.d., http://oilsands.alberta.ca/reclamation.html#JM-OilSandsArea.

[12] *Electric Power Statistics, December 2014*, StatsCanada, February 2, 2015. www.statcan.gc.ca/daily-quotidien/150226/dq150226d-eng.htm.

[13] Marjorie Cohen, "Energy Dispossession in Canada: Electricity," Institute of Public Administration of Canada Energy Conference: *Canada's Energy Security: Superpower or... a Player*, February 2008, 25–6. The next few paragraphs are derived from this.

[14] "Hydroelectricity consumption (most recent) by country," NationMaster, 2011, www.nationmaster.com. Icelanders and Norwegians use more hydroelectric power per person than Canadians.

[15] *Canada's Energy Future: Reference Case and Scenarios to 2030*, Energy Market Assessment, NEB, November 2007, 80, fig. 5.21, www.neb-one.gc.ca.

[16] *Oil & Gas Security: Canada 2010*, 4; Norway *Oil & Gas Security*, 4. These percentages come from these reports.

[17] The methane dissipates much more quickly in the atmosphere though.

[18] "Quarterly Population by Year, Provinces and Territory 2014," StatsCanada, www.statcan.gc.ca. With only 12% of Canada's population, Alberta produces a third (34%) of Canada's GHGs. Saskatchewan with 3.5% of Canada's people produces 11% of Canada's GHGs.

[19] *Report of the Canada Power Grid Task Force, Vol II,* Canadian Academy of Engineering, 2011, 23, table 1-2. Steam, internal combustion and combustion turbine, powered by coal and natural gas, account for 89% of Alberta's electrical generation.

[20] *Power Grid Task Force,* Academy of Engineering, 10.

[21] Michael Safi, "Tesla Announces Low-cost Batteries for Homes," *Guardian* (London), www.theguardian.com, May 1, 2015.

[22] Daniel Trefler, "The Loonacy of Parity: How a Strong Dollar Is Weakening Canada," *Globe and Mail,* October 16, 2010, www.theglobeandmail.com.

[23] Ibid.

[24] Kirk Makin, "Clash Over Oil Sands Inevitable: Lougheed," *Globe and Mail,* August 14, 2007, www.globeandmail.com.

[25] Makin, "Clash Inevitable."

[26] Carrie Tait, "East Coast's Hebron Offshore Play Gets the Nod from Exxon," *Globe and Mail,* January 4, 2013, www.theglobeandmail.com. Exxon-Mobil will invest $14 billion in the Hebron field, 350 km southeast of St. John's. The plan is to produce 150,000 barrels of oil a day by 2017. Hebron should reverse declines in Newfoundland's oil output.

[27] The Alberta "firewall" letter addressed to then Alberta Premier Ralph Klein as published in the *National Post,* January 24, 2001.

[28] Michael Wood, "Hands Off Our Oilsands, Stelmach Warns Party Leaders," *Herald-Tribune* (Grande Prairie), dailyheraldtribune.com;. David Staples, "Carbon Policy Must not Punish Alberta," *Edmonton Journal,* April 29, 2011, B1.

[29] The Canadian Press, "Brad Wall: NDP's Dutch-Disease Strategy Won't Fly With Saskatchewan, Premier Warns," *Huffington Post Canada,* June 7, 2012, www.huffingtonpost.ca.

[30] Keith Gerein, "Alberta Must Invest Its Energy Wealth to Prepare for Future: Economic Council," *Edmonton Journal,* May 5, 2011, A1, www.calgaryherald.com.

[31] Tanya Whyte, "Sarnia–Montreal Pipeline," (unpublished paper, March 19, 2008).

[32] Jay Hakes, *A Declaration of Energy Independence,* (Hoboken, New Jersey: John Wiley & Sons), 17.

[33] "Energy Overview, 1949–2006," US Energy Information Administration, www.eia.doe.gov.

[34] Bruce Muirhead, "From Special Relationship to Third Option: Canada, the US, and the Nixon Shock," *American Review of Canadian Studies 34,* no. 3 (2004).

[35] Whyte, "Sarnia–Montreal Pipeline."

[36] "Energy Overview, 1949–2006," US Energy Information Administration, www.eia.doe.gov.

[37] Associated Press, "5 Options Cited for North US As Canada Halts Oil Exports," *New York Times,* August 5, 1976, 56.

[38] Michael T. Klare, *Blood and Oil,* (New York: Owl Books, 2004), 113–118.

[39] Klare, *Blood and Oil,* 114.

[40] Tony Clarke, *Tar Sands Showdown,* (Toronto: Lorimer, 2008), 144.

[41] Len Flint, "Workshop Working Report," (Draft, Houston, Texas: Oil Sands Experts Group Workshop, January 24–5, 2006, Len Flint Lenef Consulting (1994), January 31, 2006).

[42] US Energy Policy Act of 2005, Public Law no. 109–58, 594 stat, 109th Congress, p. 119

[43] John Gray, "The Second Coming of Peter Lougheed," *Globe and Mail,* August 28, 2008.

[44] *Tar Sands: The Selling of Alberta,* directed by Tom Radford, White Pine Pictures and Clearwater Media, aired on CBC, March 21, 2008.

[45] Shawn McCarthy, 'US Advocates for Oil Sands Tout Security of Supply," *Globe and Mail,* October 12, 2009, B1.

[46] There is no certainty that these lines if built would send oil only to Asia. Substantial amounts might be taken by tanker to the US.

[47] Nathan Vanderklippe and Shawn McCarthy, "Without Keystone XL, Oil Sands Face Choke Point," *Globe and Mail,* June 8, 2011.

[48] Reuters, "Chu Oil Sands Are a Complicated Issue," *Business Insider,* June 1, 2009, www.businessinsider.com.

[49] "Clinton Tackled for Oilsands Comments," *CBC News,* October 21, 2011, www.cbc.ca/news/world/story/2010/10/21/clinton-senators-oilsands.html.

[50] Barack Obama, "Energy Security is National Security," speech, Governor's Ethanol Coalition, Washington, DC, February 28, 2006, obamaspeeches.com.

[51] Lappé, *EcoMind,* 146.

Index

Figures indicated by italics

Abu Dhabi, 37
Acting on Climate Change report, 185
Africa, 68
Alberta: alternative post-Sands futures, 109–10, 119–20, 122, 130–33; and eco-energy security plan, 229–30, 231–33; electricity in, 155, 228; financial assets of, 131; GHGs from, 111; and Great Recession, 131; and Hurricane Katrina, 51; 1986 recession, 49; oil and natural gas outputs, *127*; petrochemical industry in, 120; and Sands, 5; share of oil and natural gas, 149; staple theory for, 120–22; subsidization of oil from, 166. *See also* Alberta government
Alberta Energy Corporation, 133, 136
Alberta Energy Regulator, 101
Alberta government: and Big Oil, 4, 25–26; and bitumen bubble, 26; and Canadian energy strategy, 12–13, 19–21, 23, and economic rents, 126, 149; and free trade, 21–22; Heritage Fund, 126, *127,* 128; and natural gas, 95–96, 100–101; neoliberal approach of, 128; and NEP, 18, 46, 47, 49, 103; and oil royalties, 128, 129–30; and proportionality rule, 103. *See also* Alberta
Alberta Premier's Council, 109–10, 119
Alberta Treasury Branches, 133
Atlantic Canada, 180–82; energy insecurity of, 39, 52, 53, 66, 77, 78, 82, 174; energy security for, 14, 84–85, 91, 174–75, 177, 230; natural gas, 163; reliance on imports, 50, 77, 166; and SPRS, 86. *See also* Eastern Canada; New Brunswick; Newfoundland; Nova Scotia
Australia, 4, 68–69, 77, 80, 154, 198, 203

Bakken shale oil, 16, 51, 82, 159–60, 166, 173
BC Hydro, 155
Big Oil: and Alberta's royalties, 129–30; and Britain's petrol crisis, 73–74; and Canadian energy strategy, 13, 19–21, 22–23; and Canadian oil, 14, 41, 43, 44; and climate change denial, 148;

and the energy-poor, 193–94; goals of, 91–92, 204; as integrated corporations, 27; and NEP, 48; oil reserves, 63; on peak oil, 56; private vs. national companies, 67–68, 137, *138*; Seven Sisters, 67, 136-37, *138*
Bill C-45, 172–73, 231
Bill C-51, 109
bitumen, 5, 9, 172, 226. *See also* Sands
bitumen bubble, 26–27
Bolivia, 141, 145, 146, 154
BP, 19, 112, 136, 137, 147, 204
Brazil, 62, 146
Britain: BritOil, 140; energy efficiency ranking, 75; and energy security, 22, 69, 71–72, 74–75; and fuel poverty, 194; natural gas stocks, 75; oil during WWI, 9–10; oil reserves, 74; petrol famine, 53, 71–74; sufficiency campaigns in, 201; and tradable energy quotas, 213, 214, 217
British Columbia, 149, 155
Bush administration, 2, 30, 99, 236

California, 34, 112, 225
Canada, 9, 43–44, 61, 75, 104, 106, 116–19, 201, 229. *See also* Atlantic Canada; Canadian energy security; Eastern Canada; Harper government; Sands; Western Canada
Canada School of Energy and the Environment, 20
Canada-US Free Trade Agreement (FTA), 21–22, 93, 96, 97, 98, 100, 102–3, 104, *161–63. See also* NAFTA; proportionality rule
Canadian Association of Petroleum Producers (CAPP), 55, 110, 237
Canadian Council of Chief Executives (CCCE), 1–2, 20–21
Canadian dollar, 116, 229
Canadian energy security, 3–4, 12–14; additional measures for, 220; and Alberta, 12–13, 19–21, 23; approach to, 7, 14, 35, 38, 91, 171, 183–84, 221, 239; and Big Oil, 13, 19–21, 22–23, 27–28; conventional oil for, 29; energy disruptions in, 52–53; and exports to Asia, 171; federal response to, 4, 21,

50–51, 68, 69, 88, 89–90, 223; grassroots initiatives for, 185–86; history of,
18–19, 39–40, 166–67; importance of
Quebec to, 177–79; and interprovincial
electricity grids, 227–29; lack of national energy firm, 93; and natural gas,
33, 176–77; opposition to, 14, 21;
under Pierre Trudeau, 39–40, 43,
44, 46; renewables in, 218–19; and
SPRS, 15, 75–76, 78–79, 80, 81–82,
85–87; and Strait of Hormuz, 66; and
US, 14–15, 40–41, 43, 45, 78–79, 94,
233–38; vulnerability of, 4, 6–7, 12,
14, 51–52, 76, 77, 91; and west-to-east
pipelines, 28, 82–84. *See also*
eco-energy security plan; National
Energy Program, Canada; national
energy quotas; National Oil Policy
Canadian Natural Resources (CNRL), 26
cap-and-trade emissions plan, 19, 116, 211
Cape Breton Island, 53, 180
capitalism, 187–88
carbon capture storage (CCS), 112, *113*,
210
carbon emissions. *See* greenhouse gases
carbon trading, 204–5, 212
Carney, Pat, 22, 102, 103
Carson, Bruce, 20
Carter, Jimmy, 30, 33, 41–42, 234, 238
Carter Doctrine, 42–43
cheap oil, 9, 33, 60, 67, 114, 192, 196
Cheney, Dick, 36, 99, 139, 236
Cherniak, Cyndee Todgham, 92, 108
Chevron, 136, 137, 149
China: and African oil, 68; and climate
change, 101; exports proposed to,
24–25, 237; fuel efficiency in, 31, 57;
investments in Sands, 36; relocalization in, 196; resource nationalism in,
145; SPRs, 4, 80, 81, 87; and Strait of
Hormuz, 66; and US, 235; wages in,
115. *See also* CNOOC
China Syndrome, 117
Chrétien government, 104–6
climate change, 16, 101, 148, 189, 196,
208, 210–11
Climate Change Act (UK), 213
climate justice, 206
Clinton, Bill, 104, 105, 235

Clinton, Hillary, 106, 238
CNOOC, 36, 99, 145
coal-bed methane, 32, 101. *See also*
natural gas
Come by Chance refinery (North Atlantic
refinery), 52, 181
Commission on North American Freedom,
237
conservation. *See* national energy quotas;
sufficiency
Conservative Party (UK), 71, 114
conventional oil: and Canadian energy
security, 7, 14, 29, 84, 91, 167, 176,
223, 230, 233; cleaning up, 172;
export of, 35, 91; shortages of, 60–61,
98, 100, 224
co-operative paradigm, 195–96
Crown corporations, 150–51

Deepwater Horizon well, 147
de-globalization, 112, 114–15, 191, 196
Denmark, 77, 80, 88, 130, 213
deregulation, 222
diversification, 122
Dutch Disease, 115–19, 229

Eastern Canada: difficulty of transporting
oil to, 52, 171; energy insecurity of, 14,
15, 39, 51, 166–67, 180; energy security for, 81, 83–84, 171, 175, 181–82,
183; heating oil, 53, 78, 86, 181;
natural gas in, 163, 181; oil imports to,
50, 173, 175–76, 179; and pipelines to,
28, 82–83, 172; and Strait of Hormuz,
66; US shale oil into, 51, 77, 82,
159–60, 166–67, 172
easy oil: age of ending, 2, 19, 30–31,
53–62
eco-energy security plan, 76–77, 223,
225–27, 230–33, 239
ecological modernization, 204–5
economic growth, 114, 120–23, 190–91,
200, 209, 221
economic rents, 126, 148–49. *See also*
royalties
economic sovereignty, 198
eco-socialism, 200
Ecuador, 141, 146
efficiency, 188, 190–91, 206

electricity: access to, *96*; generation
efficiencies, *207*; hydroelectricity,
181–82, 184, 208, 227–28; interprovincial grids, 227–29; and public
interest ownership, 155
Enbridge, 26, 163, 164, 166, 171–72, 173.
See also Northern Gateway oil pipeline
Enbridge Line 9: and energy security, 77,
83, 166, 167; history of, 40, 43, 44, 46;
imports from, 14, 28, 82; opposition
to, 83, 171, 172; purpose of, 82–83,
233; reversal of, 51, 83, 159, 170
energy consumption, *96*
Energy East oil pipeline: capacity of, 82,
167; and eco-energy security plan,
230; and energy security, 83–84, 91,
176; McKenna on, 27; Notley on, 13;
opposition to, 19, 83, 159–60, 174,
177, 185–86; purpose of, 51, 82–83,
163, 170, 173–74
Energy Independence and Security Act
(us), 34, 236
Energy Information Administration, us
(eia), 15, 63, 235
Energy Policy Act (us), 236–37
energy poverty, 193–94
energy pyramid, *193*
energy security, "national," 13–14. *See also*
Canadian energy security
Energy Supplies Allocation Board, 50
Energy Supplies Emergency Act (can), 69
environmental nationalism, 158, 198
environmental protection, 42, 146, 198.
See also climate change
European Community, 10
European Union, 31, 74, 80, 101, 194, 211
ExxonMobil, 37, 67, 136, 137, 148, 172,
194

FERC (us Federal Energy Regulatory
Commission), 227
Fleming, David, 211–12, 213, 216
foreign ownership and control, 21, 35–38,
39, 41, 45, 48, 135–36; Keynes
criticism of, 197; and Mexico, 94–95,
142–43; and Norway, 123, *124*;
reduction of, 18, 44, 117, 154; by
transnationals and petro-elites,
13–14, 35, 93, 163, 167; vs. resource

nationalism, 139, 141–45, 157–58;
us pressure for, 43, 107
fracking, 19, 32, 95, 224; opposition to,
179, 186
France, 10, 55, 72, 144
Freedom from Dirty Oil (Price and
McEachern), 171
free trade agreements, 21–22, 92. *See also*
Canada-us Free Trade Agreement;
nafta; Security and Prosperity
Partnership of North America
fuel efficiency standards, 30–31, 42,
225–26

G5/G7/G20, 144
gasoline: prices, 192; reducing use of,
156–57; shortages, 53, 82, 180
GATT (General Agreement on Tariffs and
Trade), 97
Gaz Métro, 171, 177, 178
Gazprom, 67, 137, 171
General Motors, 109, 196
Germany, 130, 131
Gibbins, Roger, 20
globalization, 114, 144–45, 191, 196, 197
Great Recession (2008-09), 30, 89, 106,
131, 196, 209
green-energy economy, 130–33. *See also*
low-carbon economy; national energy
quotas
greenhouse gases (GHGs): from Canada,
17–18, 29, 38, 85, 92, 118–19, 199,
226–27; and climate change, 16; and
eco-energy security plan, 231;
emissions globally (2010), *17*; from fossil
fuels, *96*; from future oil
projections, 55; international agreement
needed for, 6; and interprovincial power
grids, 228; from Sands, 38, 96, 101–2,
111–12, 224; from shale gas, 32; targets
for cutting, 18, 203; and working hours,
188. *See also* cap-and-trade emissions
plan; carbon capture storage; carbon
trading
Green Jobs report (Thompson), 130–31,
132, 133
Gulf of Mexico, 19, 62, 147
Gulf wars, 33, 42–43

Harper, Stephen, 21, 23, 83, 107, 174, 223, 238
Harper government, 2, 12, 37, 109, 112, 118–19, 172–73, 229, 231
Hayden, Anders, 199, 201
Hebron oil field, 84, 149, 181
Homer-Dixon, Thomas, 132, 199, 200
Hormuz, Strait of, 65–66, 67
How Much Is Enough? (Skidelskys), 187
human rights, 192–93
Hurricanes Katrina and Rita, 51, 64–65, 98
hydroelectricity, 181–82, 184, 208, 227–28. See also electricity

Imperial Oil, 41, 181
India, 66, 80–81
indigenous peoples, 5, 6, 9, 22, 68, 83, 126, 136, 146, 149, 159–61, 170–74, 176, 182–83, 186, 201, 220
International Energy Agency (IEA): Canada in, 98; on Canadian energy security, 89; creation of, 80, 98; and 1980s oil glut, 66; on offsetting old fields, 110; and OPEC oil reserves, 63; on peak oil, 56; on Sands, 110; sharing system of, 98; and SPRs, 10, 77, 80, 88
Interprovincial Pipeline, 28, 40, 43, 44, 46, 167. See also Enbridge Line 9
Iran, 63, 65, 66, 67, 136, 140, 143
Iraq, 34, 54–55, 63, 65, 98
Irving refinery, 27, 52, 181
Islamic State of Iraq and the Levant (ISIL), 55, 65

Japan, 31, 66, 131, 189–90

Kalamazoo River (MI) spill, 172
Keystone XL pipeline, 12, 13, 23–24, 83, 100, 111, 159, 164, 172, 175
Kinder Morgan pipeline, 13, 25, 83, 159, 164, 170
Klein, Naomi, 185, 198, 204
Klein, Ralph, 18, 25, 33, 50
Kuwait, 62, 128
Kuwait Petroleum, 67, 137

Lac-Mégantic explosion, 166, 173, 222
Lambton salt caverns, 86, 87

Lappé, Frances Moore, 213, 239
Laxer, Gordon, 1, 3–4, 6
Leotta, Kathy, 68–69
Liberal Party (federal), 16, 104
Limits to Growth, The (Club of Rome), 191
liquified natural gas (LNG), 32, 170, 178–79, 224. See also natural gas
Lougheed, Peter, 35, 47, 103, 109, 119–20, 125–26, 133, 229, 237
low-carbon economy, 112, 205–6, 210–11, 213, 230. See also national energy quotas
Lunn, Gary, 50, 51

Mackenzie Valley Pipeline, 161
Malaysia, 37
manufacturing, 115, 117, 118, 119, 196, 229
Marcellus Formation (shale gas), 95, 176–77
Maugeri, Leonardo, 54, 55
May, Elizabeth, 15
Mayflower (AR) spill, 172
McDonald, Marci, 102
McKenna, Frank, 26, 27
Mexico, 48, 61, 77, 94–95, 99, 104, 140, 142–43, 235
Middle East Pipelines and Ports, 65
Montreal, 52–53
Mulcair, Thomas, 16, 116, 232
Mulroney government, 44, 49

NAFTA (North American Free Trade Agreement): and Canadian energy security, 14; chapter 11, 154; and Chrétien government, 104–6; creation of, 93, 100; and Keystone XL pipeline, 23, 100, 175; and Mexico, 94–95; and 9/11 attacks, 1; proposed Canadian action on, 98, 108, 175; purpose of, 2, 21–22; US on opting out, 106–7. See also Canada-US Free Trade Agreement; proportionality rule; Security and Prosperity Partnership of North America
National Energy Board (NEB), 13, 83, 88, 96, 223
National Energy Policy, US (NEP), 30, 90, 99, 236
National Energy Program, Canada (NEP),

1, 13, 18, 30, 40, 44, 46–49, 103, 230
national energy quotas (NEQS), 156–57,
    211–13, 214, 215, 216–19
nationalism, 142, 197–98. *See also*
    environmental nationalism; renation-
    alization; resource nationalism
National Oil Policy, 39, 43, 166–67
natural gas: in Alberta, 100–101; in
    Britain, 74–75; coal-bed methane, 32,
    101; in Eastern Canada, 163, 181; and
    energy security, 32–33, 176–77; export
    of, 32, 162–63, 168, 178–79; fleet
    vehicles switch to, 63–64; and IEA, 98;
    LNG, 32, 170, 178, 224; in Netherlands,
    115–16; and proportionality rule, 94,
    95–96; proposed plan for, 223–24,
    239; and public interest ownership,
    155–56; and Russia, 67; and Sands,
    32–33, 101, 224, 225; shale gas, 32,
    101, 178–79, 224; TransCanada main-
    line, 160–61, 163; in US, 32, 95
Natural Resources Canada (NRCan), 50,
    87, 88
Navigable Waters Protection Act (CAN),
    172–73
neoliberalism, 106, 144–45, 147–48, 197,
    205, 209–10, 222
New Brunswick, 163, 186, 228
New Democratic Party (AB), 4, 25–26,
    49, 120, 128
New Democratic Party (federal), 16, 120
Newfoundland 180–82; and Atlantic
    Canada energy security, 84–85; 91,
    175; and Big Oil, 36, 149–50, 158;
    electricity from, 180, 181–82, 183,
    228; natural gas in, 163, 181; oil
    exported from, 50, 84, 181; oil
    production in, 52, 149, 174, 180–81
Nigeria, 141, 142, 146
Nixon, Richard, 30, 234, 235
non-conventional oil. *See* fracking; Sands
Northern Gateway oil pipeline, 12–13, 25,
    26, 36, 83, 159, 170
Norway, 36, 77, 80, 88, 122–23, 124,
    125, 126, 127, 228
not-for-profit companies, 147. *See also*
    public interest ownership
Notley, Rachel, 4, 12–13, 25–26, 49, 109,
    112, 128–30

Nova Scotia, 163, 180, 182, 228
nuclear energy, 218–19, 226

Obama, Barak, 13, 22, 23–24, 30–31, 34,
    106, 107, 142, 172, 175, 238
oil, 9–10, 61–63, 114–15; and NEQS,
    156–57; ownership of, 139; as political
    weapon, 66, 67–68; volume through
    world chokepoints, 64; volume vs.
    quality, 115; Western Canada moving
    capacity, 24. *See also* peak oil; resource
    nationalism; shale oil
oil crises, 11; future results from, 64–66,
    110–11, 209–10; 1980s price crashes,
    48; preparing for, 68–69, 114;
    safeguards against, 10. *See also*
    Great Recession
oil embargoes, 65–68, 99
oil market, 67, 91–92, 93
oil reserves, official, 62–63. *See also*
    strategic petroleum reserves
oil sands, 5. *See also* Sands
Oil Shockwave report (Hirsch, Bezdek and
    Wendling),64, 65
oil spills, 146, 171, 172, 177
oil tankers, 174–75
Oil: The Next Revolution (Maugeri), 54, 55
oil wells: deep-sea, 62; horizontal drilling,
    55
Ontario, 19, 166, 172, 175, 211, 228
OPEC (Organization of Petroleum Exporting
    Countries), 62–63, 66, 80, 143, 171
Our Fair Share report, 125, 128–29

paradigms, competing, 195
Parkland Institute, 1, 2, 4, 6, 13, 22–23,
    50, 126
peak demand, 57, 60
peak oil, 16, 19, 53–54, 55–57, 56, 58–59,
    60, 61–62, 98, 196, 222–23
Pembina Institute, 204
Pennsylvania, 95, 139
Persian Gulf, 42–43
Petro-Canada, 39, 44, 48, 135, 136, 140,
    150
PetroChina, 36, 67
phase out carbon fuel exports, 5, 18,
    33–35, 84, 97, 157, 179, 219, 233–39

phase out Sands oil, 5, 23, 33–35, 92, 119, 133, 157, 225–27
pipelines: Canadian oil (c. 1977), *164*; ending need for, 176, 177; impact of routes, 170–71; lack of all-Canadian oil, 163; as nation-building, 160–61; opposition to, 159–60; proposed offshore and oil export, *169*, 170; proposed offshore natural gas, *168*; to US, 164, *165*. See also *specific pipelines*
politics and oil, 57, 60, 66, 67–68
Politics of Energy, The (Doern and Toner), 46–47
post-carbon organizations, 200
potash industry, 47–48, 158
Private Gain or Public Interest (Thompson and Newman), 148, 150, 151, 152
Progressive Conservatives (AB), 4, 5, 25
proportionality rule: and Canadian energy security, 84, 91, 97; Cherniak on, 108; and conservation, 92, 96–97; and eco-energy security plan, 231; effect of, 21–22, 77, 92–93, 97, 98–99; Lougheed on, 35; and Mexico, 77, 94–95, 104; and natural gas, 95–96; negotiating, 102–4; and NEP, 40, 49; supply basis of, 93–94; and US, 95. See also NAFTA
public interest ownership, 36–37, 135–36, 145–49, 151–58, 230–31

Quebec, 175–79; cap-and-trade system in, 211; convivial degrowth movement, 200; and electricity, 19, 180–82, 228; energy security, 14–15, 52–53, 76, 82, 84, 91, 97, 167, 176; Lac-Mégantic, 66, 176, 222; and NEP, 46; oil imports to, 50, 77, 84, 88, 95, 166, 167; 1998 Montreal ice storm, 52–53; opposition to fracking, 179; opposition to pipelines in, 170, 177, 186; pipelines to, 26, 39–40, 46, 84, 160, 167–68, 170, 175, 177–79, 230; Sands and shale oil to, 82, 166–67, 173, 176, 186; SPR for, 86–87

Rabaska LNG project, 178
railways, Canadian, 160, 173

rationing, 212–13. See also national energy quotas
Reagan, Ronald, 33, 42, 48, 66, 140
recessions, 209. See also Great Recession
Rees, Bill, 199–200
refineries, Canadian, 52, 53, 77, 81–82, 88. See also Come by Chance refinery; Irving refinery
Regional Transmission Organizations (RTOS), 227
renationalization, 196
renewable energy, 6–7, *193*, *207*, 206, 208–211, 218–20, 227–28; Britain's need for, 74; electricity as, 156, 227–28; in a low-carbon society, 29, 205, 223; and public interest firms, 150–51, 153, 156; transition to, 10, 37–38, 92, 130–33, 150–51, 182, 184. See also hydroelectricity
resource control, 46, 47
resource nationalism: and Canada, 135–36, 143, 151–54, 158; economists on, 137, 146; history of, 142–45; hybrid approach to, 154–55; nationalization of oil, 36, 39, 44, 67–68, 123–26, 136–37; negative effects of, 146; new trend towards, 141; 1970s wave, 135, 136; oil reserves controlled by, 60–61; opposition to, 139, 150; and privatization, 140–41; and public interest ownership, 157–58; reasons for, 135, 139–41; US view of, 141–42
resource pyramid, 206
royalties, 125–26, 128–130
Rubin, Jeff, 36, 114–15, 128, 203, 209
Russia, 67, 68, 136, 140, 141, 145

Sachs, Jeffrey, 203–4
salt caverns, 78, *79*, 86
Sands, 5; and alternative Albertan future, 119–20; assumptions about, 35, 50; and bitumen bubble, 26–27; and Canadian energy strategy, 19–20; and carbon capture storage, 112; and carbon emissions, 38; and China, 237; and climate change action, 101–2; and Dutch Disease, 116–19; and eco-energy security plan, 225–27; and economic linkages, 121; environmental impact

of, 29, 226–27; greenhouse gases from, 17, 92, 96, 111–12, 224; map of, 226; natural gas for, 32–33, 101, 224, 225; oil output from, 110; oil spills from, 172; opposition to, 5, 34, 110, 172; ownership of, 36–37, 135–36, 158; pipelines promoted for, 19–20, 23–25, 26, 27–28; as private investable oil, 137; sanctions against, 112; transitioning away from, 92, 133; and us, 34–35, 99, 111, 236, 237–38

Saskatchewan, 46, 47–48, 111, 149, 158, 228, 232

Saudi Arabia, 41, 57, 61–62, 63, 65, 68

Security and Prosperity Partnership of North America (spp), 1–2, 3, 4

shale gas, 32, 101, 178–179, 224. See also natural gas

shale oil, 15–16, 29, 32, 53, 57

Shell, 37, 60, 63, 67, 74, 112, 136, 137, 204

Simmons, Matt, 15, 56, 75–76, 79, 86

social responsibility, 147–48

South Korea, 37, 66, 131

Standard Oil, 41, 139

Standing Committee on International Trade, 1, 2–3

staple theory of economic growth, 120–22

Statoil, 36, 67, 112, 123, 140, 146, 150

steady-state economy, 191

strategic petroleum reserves (sprs), 10, 71, 77–78; and Canada, 15, 75–76, 78–79, 80, 81–82, 85–87; cost of, 86–87; internationally, 80–81; in salt caverns, 78, 79; strategy for building, 81–82; in us, 76, 78–79, 86

sufficiency, 186–87, 189–90, 191–92, 197, 199, 200–202

Suncor, 53, 135, 148, 150

super giant oil fields, 62

surge production, 51

tar sands, 5. See also Sands

Tasman study, 68–69

Tesla's home batteries, 228–29

Thompson, David, 130–31, 132, 133, 148, 150, 151, 152

TransCanada, 26, 82, 160–61, 163, 164. See also Energy East oil pipeline; Keystone xl pipeline

Trilateral Commission, 144

Trudeau, Justin, 16

Trudeau, Pierre, 39–40, 42, 43, 44, 46, 47

United Nations, 143–44

United States of America: and Canadian oil, 12, 40–41, 45, 94, 97, 98–99, 233–35, 236–37; and energy independence, 22, 33, 37–38, 89–90, 234, 238; energy policy, 10, 12, 29–31, 33–34, 41–43, 54–55, 99, 142, 236–37; exports to Canada, 14–15; and green economy, 131; and iea, 80; and nafta, 15; "national" energy security in, 13; nep, 30, 90, 99, 236; oil consumption, 15, 16; oil imports to, 233–34; and proportionality rule, 95, 97; and Sands, 34–35, 99, 236, 237–38; self-sufficiency, 99, 107; shale oil boom, 15–16; sprs in, 10, 76, 78–79, 86

Venezuela, 63, 67, 136, 137, 141, 143, 145, 235

Vermont Natural Resources Council, 171–72

Victor, Peter, 199, 201

Wabana iron mine, 86, 87

Wall, Brad, 116, 232

Western Accord, 102–3

Western Canada, 15, 39, 43, 50, 84

Wildrose Party, 25, 129

World Energy Outlook reports (iea), 54–57